Essentials of
ADVANCED COMPOSITE
FABRICATION & REPAIR

Essentials of
ADVANCED COMPOSITE
FABRICATION & REPAIR

Louis C. Dorworth
Ginger L. Gardiner
Greg M. Mellema

Aviation Supplies & Academics, Inc.
Newcastle, Washington

Essentials of Advanced Composite Fabrication and Repair
by Louis C. Dorworth • Ginger L. Gardiner • Greg M. Mellema
Abaris Training Staff

Published 2009 by Aviation Supplies & Academics, Inc.

See the ASA website at **www.asa2fly.com/reader/composite** for the "Reader Resources" page containing information and updates relating to this book.

Aviation Supplies & Academics, Inc.
7005 132nd Place SE • Newcastle WA 98059
Internet: www.asa2fly.com • Email: asa@asa2fly.com

Abaris Training Resources, Inc.
5401 Longley Lane, Suite 49 • Reno, NV 89511
800-638-8441 • fax 775-827-6599 • www.abaris.com

Photo credits: Cover (front) Airbus S.A.S. 2009, Albany Engineered Composites, Dexmet Corporation, Byron Moore/iStockphoto; (back) Cirrus Aircraft, ATK Space, VEC Technology. Credit and copyright information for photographs other than those owned by Abaris are acknowledged throughout the text, stated adjacent to the printed photographs. All photos not so noted are copyright Abaris 2009.

ASA-COMPOSITE
ISBN 978-1-56027-752-1

Printed in the United States of America

2019   2018   2017   9 8 7 6 5

Library of Congress Cataloging-In-Publication data:

Dorworth, Louis C.
    Essentials of advanced composite fabrication and repair / Louis C. Dorworth, Ginger L. Gardiner, Greg M. Mellema.
       p. cm.
    Includes index.
    ISBN-13: 978-1-56027-752-1 (hardcover)
    ISBN-10: 1-56027-752-1 (hardcover)
    1. Airplanes—Materials—Handbooks, manuals, etc. 2. Composite materials—Handbooks, manuals, etc. I. Gardiner, Ginger L. II. Mellema, Greg M. III. Title.
    TL699.C57D67 2010
    629.134—dc22
                              2009040060

Contents

vi

Preface

Essentials of Advanced Composite Fabrication and Repair is the companion textbook for the majority of the basic courses given by Abaris Training. It assumes no prior knowledge about advanced composite materials and chapter by chapter, progressively educates readers in using composite terminology comfortably and correctly.

After a complete description of what composites are and how they work, the book takes the reader on a comprehensive journey into the specific constituent materials involved: different types of fibers, carbon, aramid, glass, and others; matrix materials, ceramics, thermoplastic and thermoset resins and their different chemistries.

The book also goes into detail about various tooling concepts, manufacturing techniques, and accepted repair theories and concepts. Other sections include the most up-to-date information on adhesive bonding technology, core materials, and non-destructive inspection (NDI) techniques and equipment.

This book will appeal to a diverse group of readers working in the aerospace, wind energy, marine structures, sporting goods, automotive, motor sports, medical products, mass-transportation, civil engineering, and many other associated industries using advanced composite materials. It is intended to be a platform from which to present current information on a host of advanced technologies in a way that can be understood by people working at every level: from technicians to engineers, and from rank beginners to seasoned professionals.

It is the sincere desire of the authors of this book that the readers not only develop a deeper knowledge of the subject but also be empowered to use the information immediately in their workplace and career.

Acknowledgments

The authors would like to extend their genuine thanks and sincere appreciation to Mr. Michael J. Hoke, President of Abaris Training Resources, Inc. for his constant support and unremitting patience with this project; also, to Dr. Rikard Heslehurst and Mr. Greg Kress for their consummate engineering knowledge and expert input; to Mr. Dave Castellar and the worldwide Abaris Training technical staff for everything they have taught and learned over the years; and to Ms. Jonica Brunson and the Abaris administrative staff for their much-appreciated willingness to lend a hand whenever needed. And finally, to our many friends and colleagues (a list that is much too long to include them all here), without whom this book would not have been possible: to Mr. André Cocquyt, Mr. Gary Derheim, Mr. Jim Gardiner, Dr. Keith Armstrong, Dr. Scott Beckwith, Dr. A. Brent Strong, and their many, many associates and respective contacts, companies and organizations, especially those providing photographs, charts and graphs, who tolerated long questions, numerous e-mails, and after-hour phone calls in order to make this book happen.

Composite Technology Overview

Composites vs. Advanced Composites

Composites are comprised of two or more materials working together, where each constituent material retains its unique identity within the composite and contributes its own structural properties, yet upon combination the resulting material has superior properties to those of its constituents. A good example of an everyday composite material is concrete. Concrete is made with select amounts of sand, aggregate, and perhaps even glass fiber mixed with cement to bind it together. If the concrete were broken open to view inside, the individual constituents would be visible. The type and quantities of the individual materials can be adjusted to give the resulting concrete better compressive, tensile, and/or flexural properties depending on the application.

In this textbook we focus on composite laminates made from a combination of fiber reinforcement and a matrix material that binds the fibers together. Most "fiber-reinforced-matrix composites" that we see every day are made with short glass fibers mixed with a polymer or plastic matrix or resin: tubs, showers, sinks, pools, doors, fenders, and various construction materials fall into this category. (*See* Figure 1-1 on the next page.)

Highly loaded composite structures typically use continuous or long-fiber reinforcement that transfers loads along bundles or sheets (or "plies") of fibers arranged to run the length and width of the structure much like the layers in a sheet of plywood. This type of arrangement is typically used in the manufacture of structures such as: boats, bridges, snowboards, bicycle frames, race cars, and aircraft structures, to mention a few.

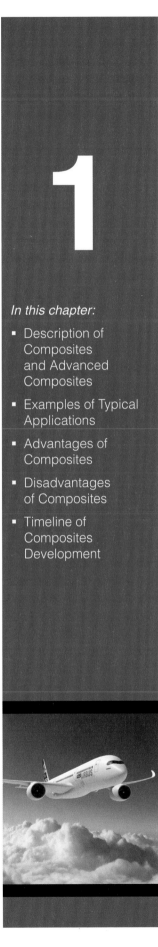

In this chapter:
- Description of Composites and Advanced Composites
- Examples of Typical Applications
- Advantages of Composites
- Disadvantages of Composites
- Timeline of Composites Development

1

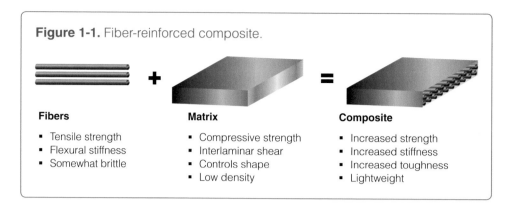

Figure 1-1. Fiber-reinforced composite.

Fibers	Matrix	Composite
• Tensile strength	• Compressive strength	• Increased strength
• Flexural stiffness	• Interlaminar shear	• Increased stiffness
• Somewhat brittle	• Controls shape	• Increased toughness
	• Low density	• Lightweight

Advanced Composites

"Advanced" composites are generally considered to be those that use advanced fiber reinforcements such as carbon fiber and Kevlar®, and that exhibit high strength-to-weight ratios. They are typically more expensive, with more precisely tailored properties to achieve a specific objective.

Fiberglass vs. Advanced Composites—Some composites are typically referred to as "fiberglass" due to their simple chopped fiberglass and polyester resin composition, whereas most aerospace parts are made using precisely laid plies of carbon fiber/epoxy prepreg. (*See* Figure 1-2.)

Examples of Typical Applications

Large components of commercial airliners—such as the Boeing 777 and the 787 Dreamliner, the Airbus A330/340, A350, and A380 aircraft, and many smaller craft such as the Bombardier Canadian Regional Jet (CRJ). (*See* Figure 1-3.)

Large primary structures on military aircraft—such as the Airbus A400 and the Boeing C-17 transports, the B-2 Spirit Stealth Bomber, V-22 Osprey tilt-rotor, as well as the F-22 Raptor and F-35 JSF fighters. (*See* Figure 1-4 on the next page.)

Many other components on modern airliners—such as radomes, control surfaces, spoilers, landing gear doors, wing-to-body fairings, and interiors. (*See* Figure 1-5.)

Large marine vessels and structures—including military and commercial vessels, as well as composite masts (the largest carbon fiber structure in the world is the *Mirabella V's* 290-foot mast). (*See* Figure 1-6.)

Primary components on helicopters—including rotor blades and rotor hubs have been made from carbon fiber and glass epoxy composites since the 1980s. Typically, composites make up 50 to 80 percent of a rotorcraft's airframe by weight. Other parts include radomes, tail cones, and large structural assemblies such as the Fenestron tail rotor on Eurocopter's Dauphin.

Figure 1-2. Fiberglass vs. advanced composites.

(Left) Chopped fiberglass and resin are sprayed onto a gel-coated mold to form the outer shell of a Class 8 truck hood. However, this is a more advanced and higher performance example of spray-up fiberglass because the shell is cured in an oven at 130°F and then reinforced with structural members made using RTM which are secondarily bonded in-place using methacrylate adhesive. (Photo courtesy of Marine Plastics Ltd.)

(Right) A technician verifies each ply during the layup process of this carbon fiber/epoxy prepreg aerospace part with an Automated Ply Verification (APV) device prior to vacuum-bagging and autoclave cure. (Photo courtesy of Assembly Guidance)

Figure 1-3.

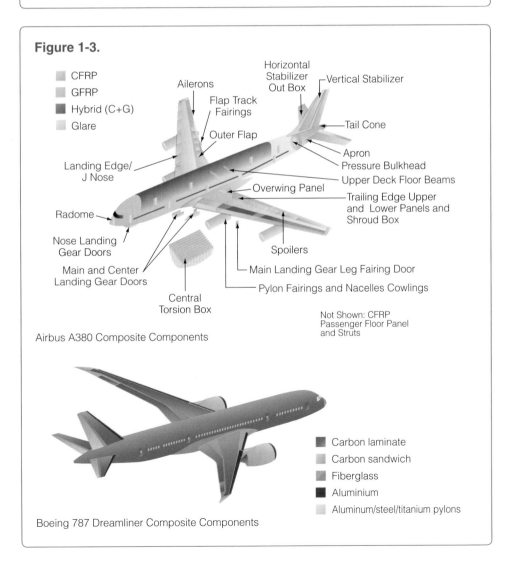

Airbus A380 Composite Components

Boeing 787 Dreamliner Composite Components

Figure 1-4.

The F-35 Lighting II Joint Strike Fighter features vertical tail feathers and horizontal stabilators made from carbon fiber-reinforced Bismaleimide composite. (Photo courtesy of Lockheed Martin)

Figure 1-5.

J-nose thermoplastic composite leading edge for the Airbus A380 made by Stork Fokker. Note the stamped thermoplastic stiffeners, which are attached using resistance welding. (Photo courtesy of Stork Fokker)

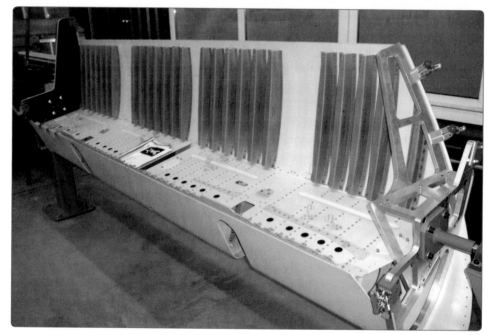

Figure 1-6.

(Left) The *Mirabella V* is the world's largest composite ship. (Photo courtesy of Select Charter Services)

(Right) The Visby class of corvettes is built using carbon fiber reinforced composite sandwich construction. (Photo courtesy of Kockums AB)

Figure 1-7.

The 429 commercial helicopter is made by Bell Helicopter Textron and features numerous advanced composite structures. (Photo courtesy of Bell Helicopter Textron)

Bell Helicopter Textron's 429 corporate/EMS/utility helicopter features composite structural sidebody panels, floor panels, bulkheads, nose skins, shroud, doors, fairings, cowlings and stabilizers, most made from carbon fiber/epoxy. (*See* Figure 1-7.)

Rotor blades for wind turbines have traditionally been made from fiberglass reinforced polyester or epoxy resins. While most manufacturers use **resin infusion** to make their wind blades, the two who use **prepreg**—Vestas and Gamesa—are also using carbon fiber in their largest 44 m/144-foot blades' spar caps. However, the industry's longest 61.5 m/200-foot blade, made by LM Glasfiber, remains fiberglass only. (*See* Figure 1-8 on the next page.)

Missiles and space vehicles rely heavily on carbon fiber composite construction both for its high strength- and stiffness-to-weight ratios and negative coefficient of thermal expansion (CTE), which gives dimensional stability in the extreme temperatures of space. The Delta family of launch vehicles have used carbon fiber and epoxy construction in more than 950 filament wound motor cases and are in their sixteenth year of production. The Pegasus rocket is the first all-composite rocket to enter service. Its payload fairing, which is 4.2 feet/1.3 meters in diameter and 14.2 feet/4.3 meters in length, is comprised of carbon/epoxy skins and an aluminum honeycomb core. The Chandra X-ray Observatory spacecraft features numerous composite structures, including the optical bench assembly, mirror support sleeves, instrument model structures and telescope thermal enclosure. Other common space structures using composites include the Mars Lander vehicle, satellite antenna reflectors, solar arrays and wave guides. (*See* Figures 1-9 and 1-10.)

Limited production sports cars are drawing from traditional Formula 1 carbon fiber composite technology and have begun to use large amounts of composites. The two-seat Porsche Carrera GT features carbon/epoxy composite (with some aramid) in the chassis, doors, doorsills, wheel wells, trunk lid, hood, underfloor, center console, and bucket seats. The Ferrari Enzo includes a carbon fiber composite monocoque chassis and driveshaft. (*See* Figure 1-11.)

Selected applications in luxury automobiles include the roof for BMW's M3 CSL Coupe, the hood and fenders for Corvette's LeMans Commemorative

Figure 1-8.

(Left) Despite its length, the LM 61.5P blade weighs only 17,740 kg/39,110 lbs. (Photo courtesy of LM Glasfiber)

(Right) Blades made for the V90 wind turbine. (Photo courtesy of Vestas Wind Systems A/S)

Figure 1-9.

ATK manufactures a wide range of spacecraft components using composites.
(Photo courtesy of ATK)

Figure 1-10.

This 60-meter mast made by ATK Space (Goleta, CA) is used to deploy an array of solar panels for use by the Space Shuttle. (Photo courtesy of ATK Space)

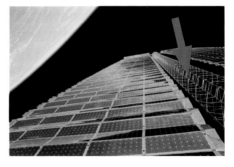

Figure 1-11.

The engine frame for the 2005 Porsche Carrera GT supercar, molded by ATR of Italy, uses carbon fiber reinforced Cycom 997 epoxy resin composite in a hot environment (engine compartment) and provides the torsional stiffness required for this open car that was previously only achievable in a closed car design. (Photo courtesy of SPE International)

Figure 1-12.

The Ford GT uses a composite rear decklid inner panel made from Torayca T600 carbon fiber (24K tow size) and epoxy resin. Molded by SPARTA Composites, it integrates components, improves dimensional control and lowers tooling cost versus stamped steel or aluminum. (Photo courtesy of SPE International)

Edition Z06, the inner deck lid and seats for the Ford GT, and the monocoque, crash structures and body panels for the 2006 Mercedes-Benz SLR McLaren. There is also a significant market in carbon fiber composite parts sold as aftermarket enhancements to production automobiles. (*See* Figures 1-12 and 1-13.)

Sporting goods such as bikes, tennis rackets and golf clubs have taken advantage of carbon fiber for decades. Carbon fiber bike frames have become commonplace, with well-known manufacturers including Kestrel, Cannondale, and Trek. Meanwhile almost all tennis rackets and a majority of golf club shafts are made using carbon fiber. A variety of hockey sticks and baseball bats feature hybrid composites using a mix of aramid, carbon and glass fiber reinforcements. Other sporting goods that commonly use aramid and carbon fibers include snow and water skis, skateboards and snowboards, fishing rods, kayaks and archery equipment. (*See* Figure 1-14.)

Medical devices continue to spur the advancement of new composites, such as ENDOLIGN™ continuous carbon fiber/PEEK thermoplastic biomaterial from Invibio Biomaterial Solutions, which is being used as an alternative to metals in the development of implantable load bearing applications such

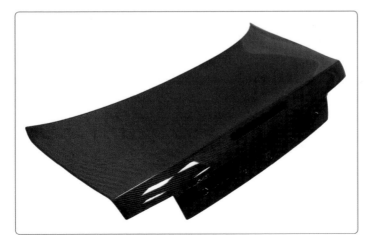

Figure 1-13.

OEM-style aftermarket carbon fiber trunk lid for the 1989-1994 Nissan 240SX 2-door made by Seibon International, Inc. (Photo courtesy of Seibon International, Inc.)

Figure 1-14.

The 2008 Cannondale Super Six is a racing oriented carbon fiber bicycle frame weighing only 920 grams/2.03 lbs. (Photo courtesy of Cannondale Bicycle Corporation)

Figure 1-15.

Invibio's ENDOLIGN composite is used in the development of orthopedic, trauma and spinal implants, such as these from icotec AG. Icotec develops and produces parts using high continuous fiber content composites and thermoplastic matrices. (Photo courtesy of Invibio)

as orthopedic, trauma and spinal implants. Other carbon fiber composites are used in medical imaging tables and accessories because they offer high stiffness and lightweight, while helping to minimize imaging issues such as signal attenuation. (*See* Figure 1-15.)

Civil Engineering structures are progressing steadily in their use of composites. Carbon fiber wraps for repair and strengthening of columns, beams, concrete slabs and bridge structures have become increasingly common. They offer an alkaline-resistant repair which is typically quicker and less costly to install due to its light weight. Carbon fiber in bridge cable stays is also progressing, offering high strength and stiffness at minimal weight as well as excellent resistance to temperature contraction and expansion due to its negative coefficient of thermal expansion. (*See* Figure 1-16.)

Figure 1-16.

The Utah Department of Transportation (UDOT) strengthened 76 columns using carbon fiber-reinforced polymer wraps. (Photo courtesy of Sika Corporation)

Advantages of Composites

High strength and stiffness-to-weight ratio. Composite structures can attain ratios 4 to 10 times better than those made from metals. However, lightweight structures are not automatic. Careful engineering is mandatory, and many tradeoffs are required to achieve truly lightweight structures.

Optimized Structures. Fibers are oriented and layers are placed in an engineered stacking sequence to carry specific loads and achieve precise structural performance. Matrix materials are chosen to meet the service environment for which the structures are subjected. (The matrix typically determines the temperature capability of the part.)

Metal fatigue is not an issue. High fatigue life is one reason composites are common in helicopter rotor blade construction, however composites do exhibit some fatigue behavior, especially around fastener and pin locations. Careful design and good process controls are required to ensure long service-life of both adhesively bonded and mechanically fastened composite joints.

Composites do not corrode. Hence their popularity within the marine industry. This is also pertinent to chemical plants, fuel storage and piping, and other applications which must withstand chemical attack.

Easily molded to shape. Composites can be formed into almost any shape, usually quite easily and without costly trade-offs in structural properties. Truly monocoque structures are possible with proper tooling.

Fewer Parts. Integrated composite structures often replace multi-part assemblies, dramatically reducing part and fastener count as well as procurement and manufacturing costs. Sometimes adhesively bonded or welded thermoplastic assemblies can almost completely eliminate fasteners, further reducing part count and production time.

Lower tooling costs. Many composites are manufactured using one-sided tools made from composites or Invar, versus more expensive multi-piece metal closed cavity, machine-tooling or large two-sided die sets that are typically required for injection molding of plastics and metal forming processes.

Aerodynamically smooth surfaces. Bonded structures offer smoother surfaces than riveted structures. Composite skins offer increased aerodynamic efficiency, whereas large, thin-skinned metallic structures may exhibit buckling between frames under load (i.e. "oil-canning").

"Low Observable" or "Stealth" characteristics. Some composite materials can absorb radar and sonar signals and thereby reduce or eliminate "observation" by electronic means. Other materials are "transparent" to radar and work well as a radar "window" in radome applications.

Disadvantages of Composites

Expensive materials. High performance composite materials and processes typically cost more than wood, metal and concrete. The cost of oil and petroleum-based raw-material products often drives the price of these materials.

Special storage and handling. Many materials such as film adhesives and prepregs typically have a limited working time (out-time), usually measured in days. Most prepregs also require frozen storage, and may have a limited shelf life of a few months to a year. Material management is critical with these materials.

Not always recyclable. Although there is some ability to recycle thermoplastic composites, recycling of composites in general is not as straight forward as with metals, wood, or unreinforced plastics. Research continues in this area.

Labor-intensive. Tailoring properties typically requires exact material placement, either through hand layup or automated processes. Each step in the process may require inspection and/or other assessment. This all requires a skilled workforce which can be expensive in some cases.

High capital equipment costs. Ovens, autoclaves, presses, controllers and software are expensive to buy and to operate. Automated machines and programming costs can be a considerable investment.

Easily damaged. Thin-skinned sandwich panels are especially susceptible to damage from low-energy impacts. Such impacts can produce delaminations which are structurally damaging yet invisible to the eye. Moisture intrusion is a special problem in honeycomb sandwich panels, which is also not apparent by visual inspection. Sometimes water intrusion can occur during manufacturing.

Special training and skills are required. A high degree of knowledge and skill is required to properly fabricate and repair advanced composite structures. Good training in this area is mandatory.

Carbon reinforcements can cause galvanic corrosion. Any metal that resides at the anodic end of the corrosion scale will corrode if in direct contact, or fastened with conductive fasteners in close proximity to carbon fiber reinforced structures.

Health and safety concerns. Health issues from working with composite materials may include dermatitis from resins, complications from inhaling respirable fibers, and exposure to suspected carcinogens in some uncured matrix systems. Other safety concerns include transport, storage and disposal of solvents and other materials classified as hazardous.

Composites Development Timeline

1932 DuPont launches the first polyamide thermoplastic polymer, nylon 6-6.

1936 DuPont is awarded patent for unsaturated polyester resin.
Fiberglas® is patented by Owens-Illinois and Corning Glass.

1937 Ray Greene makes the first fiberglass/polyester sailboat.

1938 P. Castan and S. Greenlee first patent epoxy resin chemistry.

1942 Owens-Corning begins making fiberglass/polyester parts for WWII aircraft.
Ray Greene produces a fiberglass/polyester composite day sailer.

1944 First plane with a GFRP fuselage flown at Wright-Patterson Air Force Base.
Allied forces land at Normandy in ships made of GFRP components.

1945 Over 7 million pounds of fiberglass shipped for primarily military applications.
Hartzite, a proprietary composite material, is developed and used in the construction of propeller blades by the Hartzell Company.

1946 Owens-Corning produces fiberglass reinforced plastic fishing rods, serving trays and pleasure boats.
Ciba commercializes the first epoxy resin.

1947 U.S. Navy issues contract to R.E. Young and M.W. Kellogg for filament winding machinery to manufacture fiberglass-polyester rocket motors and FRP pipe.

1948 Several thousand commercial FRP pleasure boats are produced.
Glastic Corporation develops sheet molding compound (SMC) and bulk molding compound (BMC) type materials for making automotive parts.
FRP pipe is introduced.

1950 Goldsworthy develops pultrusion composites manufacturing process.
Muskat patents Marco method for liquid resin infusion of FRP parts.
FRP composites are used for equipment in the chemical, pulp and paper, waste treatment, and other industries for corrosion resistance.

1953 The Chevy Corvette is the first automobile with a body made entirely from fiberglass reinforced plastic.

1957 Scientists invent carbon fibers from cotton and rayon.

1958 Saint-Gobain produces fiberglass composite helicopter blades for the Alouette II in their Chambery, France factory.

1959 Smith patents process for vacuum molding large FRP structures.

1960 Nancy Layman creates the first semi-permanent polymer release agent.

1961 Expanded hexagonal paper cores are used in building panels.
Carbon fibers are first manufactured from polyacrylonitrile (PAN).

1964 H-301 Libelle ("Dragonfly") non-powered glider is the first all-fiberglass composite aircraft to be issued a type certificate (in U.S. and Germany).
Geringer receives patents for Resin Transfer Molding (RTM) process.

1969 Rigid foams are used as core materials in building panels.
Boron-epoxy rudders are installed on an F-4 jet made by General Dynamics.
Windecker AC-7 is the first composite aircraft to receive FAA certification.

1971 DuPont releases Kevlar® aramid fiber.
Small amounts of PAN carbon fiber is being sold to industry.
Hexcel develops the first production ski with a honeycomb core.

1973	Development work begins on carbon fiber fishing poles and golf club shafts. The first fiberglass/foam core water ski is developed.
1974	Lockheed delivers C-130 transporter with boron-epoxy composite wing to U.S. Air Force.
1975	The first carbon-tubed, metal-lugged bicycle frame is built by Exxon Graftek but suffers frequent failures.
1976	Modified Vought A-7D Corsair II attack bomber, with graphite, boron, and epoxy composite wing, flies.
1977	Bill Lear designs the Lear Fan, the first all carbon fiber-epoxy airframe. The aircraft never made it to certification, however almost every composite aircraft design today owes something to the technology developed and pioneered at Lear Fan.
1978	Hartzell reintroduces composite propeller blades, the industry's first structural advanced composite blade for the CASA 212 utility aircraft.
1981	John Barnard develops, and Hercules Aerospace builds the first carbon fiber monocoque chassis for the McLaren Formula-One racing team.
1982	Five Boeing 737-200 commercial airliners are placed into service with composite horizontal stabilizers.
1985	Carbon fiber bicycle frames are produced in quantity by Trek. Airbus A310 is the first commercial aircraft to feature a carbon fiber/epoxy composite vertical fin (vertical stabilizer) torque box.
1986	Resin film infusion process and apparatus is patented. World's first highway bridge using composites reinforcing tendons is built in Germany. Voyager carbon fiber/epoxy composite aircraft built by the Rutan Aircraft Company/Voyager Aircraft; it made the first nonstop flight around the world without refueling.
1987	350 composite components, mostly secondary structures, entered into commercial airline flight service.
1988	The Beech Starship is the first all-composite engine-powered aircraft to receive FAA certification with a carbon fiber fuselage and wing. Airbus A320 is the first commercial aircraft to enter service with a complete composite vertical tail.
1989	B-2 bomber flies, featuring all-composite carbon fiber skins and structures, the first aircraft to use composites so extensively. Goode Skis invents the first carbon fiber snow ski pole.
1990	Bill Seemann patents Seemann Composites Resin Infusion Molding Process (SCRIMP).
1992	First all-composites pedestrian bridge installed in Aberfeldy, Scotland.
1994	Goode Skis develops the first carbon fiber water ski.
1995	Boeing 777 entered service, boasting the first all-composite empennage on a commercial jet (complete tail assembly including horizontal and vertical stabilizers, elevators, and rudder).
1996	First FRP reinforced concrete bridge deck was built at McKinleyville, WV, followed by the first all-composite vehicular bridge deck in Russell, KS.
2002	Airbus A340 enters service as the first aircraft to replace aluminum with thermoplastic composites on wing components, using fiberglass-reinforced PPS j-nose leading edge structures made by Stork Fokker AESP.
2005	First flight of the Airbus A380, which is the first large jet airliner to have a carbon fiber composite central wingbox.

Matrix Technology

2

In this chapter:
- Matrix Systems Overview
- Thermosets
- Thermoplastics
- Other Matrix Materials
- Liquid Resins
- Prepregs

Matrix Systems Overview

The matrix acts to bond and/or encapsulate the fibers, enabling the transfer of loads from fiber to fiber. It also moderately protects the fibers from degradation due to environmental effects, including: moisture, ultraviolet (UV) radiation, chemical attack, abrasion and impacts. Matrix materials can be molded, cast, or formed to shape. Types include: polymeric (plastic), metallic, and ceramic. Matrix-dominated structural properties include compression, interlaminar shear, and ultimate service temperature.

Selection of a matrix material has a major influence on the shear properties of a composite laminate, including interlaminar shear and in-plane shear. The interlaminar shear strength is important for structures functioning under bending loads, whereas the in-plane shear strength is important under torsion loads. The matrix also provides resistance to fiber buckling in a laminate under compression loads and therefore is considered a major factor in the compressive strength of a composite.

Thermoset resins are primarily used for highly loaded structures because of their high strength and relative ease of processing. Thermoplastic resins are utilized where toughness or impact resistance is desired, or when high-volume production dictates the need for a fast processing material. Metallic and ceramic matrices such as titanium and carbon are primarily considered for very high temperature applications (> 650°F/343°C).

Thermosets

With thermoset resins, the molecules are chemically reacted and joined together by **cross-linking**, forming a rigid, three-dimensional network structure. Once these cross-links are formed during cure, the molecules become locked-in and cannot be melted or reshaped again by the application of heat and pressure. However, when a thermoset has an exceptionally low number of cross-links, it may still be possible to soften it at elevated temperatures.

Thermoset resins typically require time to fully react or "cure" at temperatures ranging from room temperature to upwards of 650°F (343°C), depending on the chemistry. Examples of thermoset matrix materials include: polyester, vinyl ester, polyurethane, epoxy, phenolic, cyanate ester, bismaleimide (BMI), and polyimide resins.

Many thermoset resins bond well to fibers and to other materials, for this reason, many thermoset resins are also used as structural adhesives, paints, and coatings. Structural properties for some common thermoset matrix systems at 77°F (25°C) are shown in Tables 2.1 and 2.2 below.

TABLE 2.1 Typical Thermoset Matrix Systems for Composites

Matrix System	Tensile Strength Ksi (MPa)	Tensile Modulus Msi (GPa)	% Elongation to Failure	Cost
Polyester	3–11 (20.7–75.8)	.41–.50 (2.8–3.4)	1–5	Low
Vinyl Ester	10–12 (68.9–82.7)	.49–.56 (3.4–3.9)	3–12	Low-Med
Epoxy	7–13 (48.3–89.6)	.39–.54 (2.7–3.7)	2–9	Medium
Phenolic	2–9 (13.8–62.1)	.60 (4.1)	1–2	Medium
Bismaleimide (BMI)	7–13 (48.3–89.6)	.48–.62 (3.3–4.3)	1–3	High
Cyanate Ester (CE)	7–13 (48.3–89.6)	.40–.50 (2.8–3.4)	2–4	Very High
Polyimide (PI)	5–17 (34.5–117.2)	.20–.70 (1.4–4.8)	1–4	Very High

TABLE 2.2 Initial Cure Temperature vs. Service Temperature for Thermosets

Matrix System	Initial Cure Temperature	Operational Service Temperature*
Polyester	R/T–250°F R/T–121°C	135°–285°F 58°–140°C
Vinyl Ester	R/T–200°F R/T–93°C	120°–320°F 49°–160°C
Epoxy	R/T–350°F R/T–177°C	120°–360°F 49°–182°C
Phenolic	140°–450°F 60°–232°C	300°–500°F 148°–260°C
Bismaleimide (BMI)	375°–550°F 190°–288°C	400°–540°F 204°–282°C
Cyanate Ester (CE)	250°–350°F 121°–177°C	200°–600°F 93°–316°C
Polyimide (PI)	640°–750°F 316°–399°C	500°–600°F 260°–316°C

*Note: The operational service temperature of any thermoset resin will largely depend upon the ultimate Glass Transition Temperature (T_g) of a specified resin chemistry, as well as the cure/post-cure time and temperature that the resin has seen during processing.

Glass Transition Temperature (T_g) and Service Temperature

The glass transition temperature is the temperature at which increased molecular mobility results in significant changes in the properties of a solid polymeric resin or fiber. In this case, the upper temperature "glass transition" (or T_g) refers to the transition in behavior from rigid to "rubbery."

This can be thought of as the temperature above which the mechanical properties of a cured thermoset polymer are diminished. While it is not necessarily harmful for a structure to see temperatures moderately above the T_g, the structure should always be supported above this temperature to prevent laminate distortion.

When curing a thermoset polymer, the "rate of cure" (or rate of chemical reaction) is accomplished faster above the glass transition temperature than below it. Therefore, final cure temperatures are typically engineered to be as near as practical to the final desired T_g in order to minimize the cure time.

During processing it is important to differentiate between the "state-of-cure" (percent of chemical reaction completed) and the glass transition temperature. A common misconception is that a selected cure temperature alone determines the final glass transition temperature of the polymer. Take the case of a *partially reacted* thermoset polymer. It will continue to "cure" over time at a given temperature until it is completely reacted, which can actually elevate the T_g. For example, a thermoset polymer that is heated to 350°F (177°C) may initially have a T_g at or below this temperature. However, after several hours at this temperature, it may ultimately attain a T_g in the range of 370°–390°F (188°–199°C). Therefore, just attaining a specific cure temperature for a limited time will not necessarily produce the desired ultimate glass transition temperature of the polymer.

A similar concern is that a material may appear to be properly cured because it has been to a specified temperature and it exhibits the required strength and stiffness properties when tested at room temperature, but it may have not yet achieved a full state of cure nor the ultimate T_g required to carry design loads under "hot/wet" service conditions.

Alternatively, a *fully reacted* polymer with an ultimate T_g of, for example, 275°F (135°C), will not necessarily increase regardless of added temperature or time at that temperature. *See* Figure 2-1 (on the next page).

After initially processing a polymer to a full state of cure, the resulting T_g is typically referred to as the "dry" T_g. Ingress of moisture or other fluids into the structure when exposed to a hot-wet service environment reduces the T_g. This is often referred to as the "wet" T_g and it is considered to be the upper temperature limitation of the polymer. As a rule, the maximum designed service temperature limit of a composite structure is usually conservatively below this threshold.

The T_g of a given **neat-resin** or composite sample can be established using the following methods of Thermal Analysis (TA) shown in Figure 2-1: Thermo Mechanical Analysis (TMA), Dynamic Mechanical Analysis (DMA), and Differential Scanning Calorimetry (DSC).

Figure 2-1. Example of a post-process DMA to establish a T_g.

Tested on a parallel plate rheometer to approximately 350°F (177°C). The purple trace (Gp) is measuring the modulus of the sample. As the sample is heated, it exceeds its T_g and the modulus drops rapidly. The onset of this slope is approximately 275°F (135°C) according to the graph. (Rheometer, DSC, process control equipment, software, and all thermal analysis data/graphs courtesy of AvPro, Inc., Norman, OK in collaboration with Abaris Training Resources, Reno, NV)

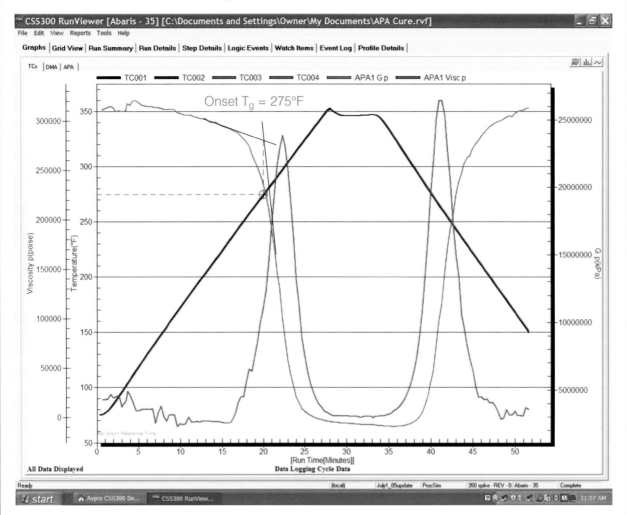

Common Thermoset Matrix Systems

Polyester

Unsaturated polyester resins typically achieve cross-linking through a multi-phase reaction. They are inexpensive and provide fairly good environmental resistance. Polyesters have lower strength and exhibit higher shrinkage than epoxies during cure. They generally contain between 30–50% styrene in the formula, producing considerable emissions.

Vinyl Ester

Vinyl esters have an epoxy backbone with vinyl groups at the end of the molecular chain, connected by ester linkages. Thus, they achieve cross-linking through a multi-phase reaction, similar to polyester, and yet offer better environmental resistance, strength and fiber adhesion than polyester resins due to their epoxy components. For this reason, they are often called "epoxy vinyl esters." They exhibit less shrinkage than polyester, but more than epoxy. Their cost is also in between polyester and epoxy, while their styrene content and emissions are roughly the same as polyester resins.

Epoxy

The workhorse of the advanced composites industry, epoxy resins offer better physical and adhesion properties than polyester or vinyl ester. Epoxies cure by cross-linking reactive polymers at both room and elevated temperatures. They provide good environmental resistance, but emit large quantities of toxic smoke when burned. More expensive than polyester and vinyl ester, epoxies are available with a wide variety of thermal and structural properties.

Phenolic

Phenolic resins are simply reacted phenols and aldehydes (phenol formaldehyde resin). They are fairly brittle, with poor adhesive characteristics, but have excellent **static-dissipative** properties and exhibit good chemical resistance. Stable at slightly higher temperatures than epoxies, phenolics offer good flame resistance and low smoke toxicity. They are also used for their ablative properties.

Cyanate Ester

Cyanate esters offer an excellent balance of mechanical performance and toughness. They are very sensitive to moisture uptake prior to processing, but provide high temperature service capabilities after post-cure. They also exhibit minimal microcracking and low moisture absorption after cure. Cyanate ester resins have a low coefficient of thermal expansion (CTE) compared to epoxy. They provide good dielectric properties and are somewhat expensive.

Bismaleimide

Often referred to as BMI, these resins deliver better thermal stability than epoxies with comparable processing. They typically have very low viscosity when processed and can be brittle, although improved toughness formulations are available. BMI resins exhibit good hot/wet properties, and are typically more expensive than epoxies.

Polyimide

Polyimides exist in both thermoplastic and thermoset formulations. They require a high temperature cure (640–750°F/316–399°C) and can be difficult to process. They provide good flame resistance and low smoke-toxicity, but are fairly expensive.

Principles of Curing and Cross-linking

Initiated Systems

Polyester molecules want to "polymerize" together into long chains. Because of this, the resin manufacturers add "inhibitors" which prevent polymerization. Later, an initiator is used to consume these inhibitors, thus allowing the chain-polymerization to proceed. These molecular chains are then cross-linked together with a cross-linking agent (usually styrene) in a second phase of reaction during the cure.

Most of these systems contain up to 50% styrene (or methacrylate or vinyl toluene) as a functional monomer. This typically leads to a strong styrene odor during mixing and curing, which may be objectionable. In addition, there are often limits imposed by regulatory authorities on airborne styrene concentrations. "Closed molding" processes are now typically prescribed for use with these systems to effectively reduce the airborne styrene emissions.

Other items, such as promoters and accelerators, are added to these resin mixes to influence the rate of cure. Cobalt Napthenate (CoNap) is a common *promoter*, and Dimethylaniline (DMA) is an *accelerator*.

Technically not a catalyst, but often referred to as such, the *initiator* starts the reaction that allows the chemistry to evolve to a cured solid. The most commonly used materials for this purpose are:

- Methyl Ethyl Ketone Peroxide (MEKP)
- Benzoyl Peroxide (BPO)

One must be particularly careful not to mix MEKP or BPO and CoNap directly together, because they will violently react with each other, and catch on fire or explode, unless the CoNap is diluted into the resin first.

Normally, $0.5-2.0\%$ by weight of the initiator is used to facilitate the reaction, depending on the ambient temperature. Within limits, one can control the speed of polymerization of these systems by varying the amount of initiator. Mixing "hot" (extra initiator) causes the reaction to happen faster and adding less slows it down ("cold").

Epoxy Systems

Epoxies are cured by mixing the resin with a "hardener," not an "initiator." Most epoxy resins are derived from bisphenol "A" or "F" (Novalac) materials. Bisphenol A or F is reacted with epichlorohydrin to form "diglycidyl ether of bisphenol A" (or F) (DGEBA or DGEBF). This is the base resin in a typical two-part epoxy system. Several amine and anhydride curing agents are used to cross-link with epoxy resins depending on the desired end-properties:

- **Amines** are basically ammonia with one or more hydrogen atoms replaced by organic groups. Aliphatic amines and polyamines are common curing agents for epoxy resins.
- **Cycloaliphatic amines** (Carbon-ring structured aliphatics). These hardeners provide better moisture and UV resistance and are often prescribed

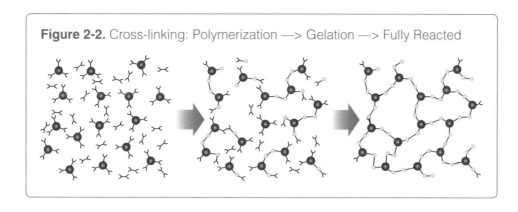

Figure 2-2. Cross-linking: Polymerization —> Gelation —> Fully Reacted

when moisture is an issue in handling or processing the epoxy resin.

- **Amides and polyamides.** Amides are basically ammonia with a hydrogen atom replaced by a carbon/oxygen and organic group.

- **Anhydride Curing Agents** such as Dodecenyl Succinic Anhydride (DDSA), Methyl Hexahydro Phthalic Anhydride (MHHPA) Nadic Methyl Anhydride (NMA), and Hexahydrophthalic Anhydride (HHPA), provide for good electrical and structural properties and exhibit a long working time. Anhydrides require an elevated temperature cure.

Amine-based curing agents are more durable and chemical resistant than amide-based curing agents but most have a tendency to react with water or moisture in the air. Amides, on the other hand, are more tolerant and less affected by moisture.

Aliphatic amines generally have a short reaction time (short pot life) and aromatic amines a longer reaction time (longer pot life). In these systems, the base resin molecules attach to the hardener molecules when mixed. These polymers cross-link, or form strong chemical links to each other.

Since the resin and hardener molecules are cross-linked to each other to make one large molecular structure, it is of utmost importance that the two materials be mixed at the proper ratio. Unlike polyesters and vinyl esters, one cannot mix these systems "hot" or "cold." They may never obtain the desired structural or thermal properties if mixed incorrectly. These systems typically go through a change from a liquid to a gel to a solid, as illustrated above in Figure 2-2.

The time it takes for a thermoset resin to gel at a given temperature is an important marker in processing these resins. All flow and compaction steps must be accomplished prior to this point. Figure 2-2 illustrates the various stages of a thermoset resin as it cross-links.

Thermoplastics

Thermoplastic matrix materials undergo a physical change from a solid to a liquid when heated, and then re-solidify upon cooling. This is because thermoplastic polymers have linear or branch-chain molecular structures, with no chemical links between them. They are instead held in place by weak intermolecular bonds (Van der Waals forces).

Because thermoplastic molecules are not cross-linked, the intermolecular bonds in a thermoplastic polymer can be broken when heated, allowing the molecules to flow and move around. Upon cooling, the molecules freeze in their new positions, restoring the weak bonds between them, thus returning to a solid form. Theoretically, a thermoplastic polymer can be heated, softened, melted, and reshaped as many times as desired. Practically, this may cause aging of the matrix and a change in properties.

Depending on the chemistry, thermoplastics typically require temperatures in excess of 300° (149°C) before they start to melt or flow. Some engineering thermoplastics require temperatures in excess of 650°F (343°C) to achieve flow.

Examples of thermoplastics include: polycarbonate (PC), Polyamide (PA), polyethylene (PE), polyethylene terephthalate (PET), polyetheretherketone (PEEK), polyetherketoneketone (PEKK), polysulfone (PSU), polyamideimide (PAI), polyphenylene sulfide (PPS), and many others. Table 2.3 below is a partial list of typical engineering thermoplastics used in advanced composite structures.

TABLE 2.3 Typical Thermoplastic Matrix Systems for Composites

Matrix Description	Process Temperature	Maximum Service Temperature*
Polyetherketoneketone (PEKK)	620–680° F 327–360° C	300° F 149° C
Polyetherimide (PEI)	625–675° F 329–357° C	410° F 210° C
Polyphenylene Sulfide (PPS)	610–650° F 321–343° C	285° F 140° C
Polyetheretherketone (PEEK)	725–755° F 385–413° C	340° F 171° C
Polyimide (PI)	540–750° F 280–399° C	600° F 316° C
Polyamideimide (PAI)	540–650° F 280–343° C	500° F 260° C

* Based on maximum Heat Distortion Temperature (HDT) listed for each chemical group.

Advantages

- Faster processing—cycle times of minutes instead of hours
- Re-flow capability—possible recycling
- Room temperature storage
- Minimal out-gassing
- Greater damage tolerance with some systems
- Possibility of fast, easy temporary repairs
- High resistance to delamination
- Typically low moisture uptake and low microcracking

Disadvantages

- Requires high to very-high temperature processing and often high pressure.
- Complex shapes can be difficult due to lack of **tack** and **drape** in prepregs.
- Expensive tooling
- Limited design database
- Woven fabric prepregs are expensive due to difficulty of impregnation — an alternative may be biaxial or multi-axial stitched forms.

Other Matrix Materials

Metal Matrix

Metal matrix composites (MMC) have a metallic, rather than an organic, matrix system. Metals such as aluminum, titanium, copper, and magnesium are used, with aluminum and titanium being the most common. Fiber reinforcements include carbon, silicon carbide, boron, aluminum oxide, tungsten, and other metallic and ceramic fibers.

Metal matrix composites can be very strong, and exhibit excellent high-temperature properties. However, they tend to be quite expensive and difficult to process. They are largely found in aircraft and spacecraft applications. Examples of metal matrices include:

- Aluminum and AL alloys
- Titanium alloys
- Magnesium alloys
- Copper-based alloys
- Nickel-based alloys
- Stainless steel

Ceramic Matrix

Ceramic matrices are used for very high-temperature applications (up to 2,000°F/1,093°C). They are brittle materials and expensive to manufacture compared to more common thermoset matrices. Examples include:

- Aluminum oxide
- Carbon
- Silicon carbide
- Silicon nitride

Ceramic matrix composites (CMC) attempt to make use of the excellent high-temperature properties of ceramics, while decreasing their brittleness and increasing their strength through the use of fiber reinforcements. Two major forms of fiber reinforcements are used:

- Short lengths of reinforcement fibers randomly distributed within the ceramic matrix;
- Continuous fiber reinforcements with high fiber content.

The former has considerably greater fracture toughness than unreinforced ceramics, while the latter gives a higher temperature capability than do metal matrix or organic matrix systems.

Carbon Matrix

Carbon used as a matrix material is **amorphous carbon**. It is usually reinforced with carbon fiber, and is referred to as "carbon-carbon." Typical applications include brakes, clutches, and high-temperature aerodynamic surfaces, such as the nose cone and wing leading edges of the space shuttle. Carbon-carbon (C-C) brake disks are considerably lighter than steel disks, and due to the material's high coefficient of friction at high temperatures, the hotter the brakes get, the better they work, unlike the brake fade suffered by steel brakes. Carbon-carbon composites are made by building up a carbon matrix around carbon fibers. There are two common ways to do this: chemical vapor deposition and resin pyrolization.

Chemical Vapor Deposition

Chemical vapor deposition (CVD) begins with a **preform**, usually made from several plies of woven carbon fabric, that has already been formed into the desired shape of the part. The preform is heated in a furnace pressurized with an organic gas, such as methane, acetylene or benzene, and then under high heat and pressure, the gas decomposes and deposits a layer of carbon onto the preform fibers. The gas must diffuse through the entire preform for the matrix to be uniform; thus, the process is very slow, often requiring several weeks and several processing steps to make a single part.

Resin Pyrolization

Another method is to use a thermosetting resin such as epoxy or phenolic, and apply it under pressure to the preform, which is then pyrolized into carbon at high temperature. A preform can also be made from resin-impregnated carbon reinforcements, and then cured and pyrolized. Shrinkage in the resin during carbonization results in tiny cracks in the matrix and reduced density, so that the part must then be re-injected and pyrolized up to a dozen times in order to fill in the small cracks and achieve the desired density. CVD may also be used to complete densification.

Carbon-carbon composites have very low thermal expansion coefficients, making them dimensionally stable at a wide range of temperatures, and they retain their mechanical properties even at temperatures above 2,000°C (3,632°F) in non-oxidizing atmospheres. They also have high thermal conductivity but resist thermal shock, meaning they won't fracture even when subjected to rapid and extreme temperature changes. Their mechanical properties vary depending on the fiber type, fiber fraction, textile weave and matrix precursor properties. Modulus can be very high, from $15-20$ GPa for composites made with a 3D fiber felt preform to $150-200$ GPa for those made with unidirectional fibers. Other properties include lightweight, high abrasion resistance, high electrical conductivity, low moisture absorption, and resistance to biological rejection and chemical corrosion (chemical attack).

Carbon-carbon composites are actually low in strength and the matrix is very brittle. They oxidize readily at temperatures between $600-700°C$ $(1,112-1,292°F)$ in the presence of oxygen. To prevent this, a protective coating (usually silicon carbide) must be applied, adding an extra manufacturing step and cost to the production process, which is already very expensive. Because of this, and the complexity of their manufacture, carbon-carbon composites have been mostly limited to aerospace, defense and some Formula 1 racing applications.

Liquid Resins

Introduction to Laminating Resins

Liquid laminating resins typically have a fairly low viscosity at room temperature (< 60 poise/6,000 centipoise at 77°F/25°C). This allows for easy fiber **wet-out** in a **layup** or an **infusion** operation. In general, the lower the viscosity, the better the flow and fiber permeability.

Structural laminating resins are usually unfilled or lightly filled with select fiber, mineral, or metallic fillers that aid in the retention of the matrix between the fibers and/or provide resiliency. A variety of modifiers may also be used in laminating resins to improve toughness properties. These modifiers range from reactive liquid elastomers (butadiene-acrylonitrile) to particulate thermoplastics.

Pot Life, Working Time, and Open Time

Mixed reactive resins and adhesives have a designated amount of time in which to use the material before it eventually becomes gelatinous and unusable in the cup. The amount of time-to-gel in the cup is usually listed on the material's technical data sheet and is specified as the "pot life." The pot life refers to the time that a specific amount of mixed resin takes to gel at room temperature (77°F/25°C) in a standardized container.

The "working time" of a resin may be quite different than its pot life. The working time is measured after the resin is mixed and distributed into a thinner cross-section (smaller mass) within the laminate or along the bondline, thus allowing more time for vacuum bagging or other processing before the resin gels. Note: The working time can be greatly affected by the amount of time that the mixed resin is left in the cup or bucket prior to use. It is always suggested that the bucket time be minimized and the resin distributed quickly to allow for maximum working time.

"Open time" refers to the time that a mixed resin or adhesive is open and exposed to the air. Open time can be an issue with certain amine-cured epoxy systems that take on CO_2 and H_2O from the air and form a carbonate layer (hydrates of amine carbonate). This layer can greatly affect the bond strength of the resin or adhesive at the interface and steps should be taken to minimize such exposure.

Mix Ratios

More often than not, when a technician is required to use a two-part resin or adhesive system, he or she will refer to the mixing instructions and determine if the mix is to be parts by weight (P.B.W.), or parts by volume (P.B.V.).

For example, at a mix ratio of 100:10 P.B.W., the technician could simply mix 100 grams of resin to 10 grams of hardener, for a total of 110 grams of mixed resin. (Or, 50 grams to 5 grams for a total of 55 grams.) They would then use what is needed and dispose of the rest. Although this method generates considerable waste, it is easy to figure out, and maintains the proper ratio of resin to hardener, thus ensuring good resin properties.

When a larger quantity is desired, it may be necessary to calculate the total amount of mixed resin that will be required. For example, say that the technician needs a total amount of 340 grams of mixed resin. For this calculation, it may be convenient to convert the ratio to a percentage and do the math to find out exactly how many grams of resin and hardener will be weighed to mix the required amount at the correct ratio.

In order to proceed, it is important to understand the difference between a percentage and a ratio. A percentage is by definition, a portion of one hundred. A ratio is used to express the relationship of two quantities.

The following example may be used as a guideline to properly convert a ratio to a percentage and then figure out the net amount of resin to hardener that will be needed to make a 340 gram total mixed batch:

Mix ratio..100:10 P.B.W

Add the two amounts to get the sum ....100 + 10 = 110

Divide each side of ratio by the sum ....100 ÷ 110 = .91
and   10 ÷ 110 = .09

(Result is percent resin and percent hardener; 91% and 9% respectively.)

Now, plug in the total amount required: ....340 x .91 = 309.4
and   340 x .09 = 30.6

The final weight of each material would be 309.4 grams of resin plus 30.6 grams of hardener to equal the 340 gram total mixed weight required.

Understanding Viscosity

The common unit for expressing absolute viscosity is "centipoise" (1/100 Poise, 1/1,000 of a Poiseuille.) The name *Poiseuille* (pronounced pwäz-'wē) and the shortened form, Poise, come from the French physician, Jean Louis Poiseuille, who performed numerous tests on the flow resistance of liquids through capillary tubes and published a paper on his research in 1846.

Poise (P) is a measure of resistance to flow and is expressed in units of one dyne-second per cm^2. Industry uses the centi-Poise (cP) or one-hundredth of a Poise because water has a viscosity of 1.002 cP (which is very close to 1 cP).

The lower the Poise or centipoise value, the more easily a material will flow. If a fluid has a high viscosity it strongly resists flow. If a fluid has a low viscosity, it offers less resistance to flow. The viscosities of liquids are greatly affected by temperature. Fluids exhibit a lower viscosity when heated, while the viscosity of gas rises with temperature.

TABLE 2.4 Viscosity Comparisons of Some Common Materials

Material (at 68°F/20°C)	Viscosity (cP)
Acetone	0.3
Water	1.002
Ethylene Glycol	19.9
Huntsman RenInfusion® 8601/8602	325
Huntsman Epocast® 50A/946 Lam. Resin	2,400
Pancake Syrup	2,500
Huntsman Epocast® 52 A/B Repair Resin	5,500
Chocolate Syrup	25,000
Ketchup	50,000
Loctite Hysol EA 9394 Paste Adhesive	60,000
Peanut Butter	250,000
Tar Or Pitch	3×10^{10}

Storage and Shelf Life Considerations

Base resins typically store well at room temperature, usually one to three years depending on the product. Some systems may require cold storage (typically $< 40°F/4.4°C$). However, they must be kept moisture-free. Because of this, many manufacturers state that the shelf life is assured as long as the product is unopened. "Pre-promoted" polyesters and vinyl esters may have shorter shelf lives, on the order of three to six months. Catalysts, hardeners, accelerators, etc. also may have short shelf lives at room temperature, on the order of several weeks to a year.

Curing Considerations

It is misleading to think that a resin has fully cured just because it appears to be hard. Most "room-temperature-curing" resins and adhesives will develop a partial cure (polymerized, partially reacted, or "gelled") within a few hours. Complete reaction and full cured properties may require several days or weeks at room temperature (77°F/25°C). An elevated temperature post-cure is sometimes prescribed for these resins to shorten the cure time and/or to raise the T_g.

Other thermoset chemistries require heat in order to facilitate active molecular cross-linking within a reasonable time, and thus will require an elevated temperature cycle to make this happen. These materials are generally specified for more high-performance applications and are usually processed in a similar manner to prepregs. (*See* **Curing Prepregs**, Page 31.)

Prepregs

Many of the disadvantages encountered with wet layup can be overcome with the use of prepregs. A "prepreg" is a reinforcement material that is pre-impregnated with resin. The reinforcement can be made from any type of fiber, in a unidirectional, stitched, or woven form. With thermosets, the resin is already mixed prior to saturating the reinforcement. The amount of resin in the prepreg is strictly controlled to achieve the desired resin content.

Prepregs can be purchased from numerous sources. Companies that produce prepregs are commonly called "prepreggers." These companies specialize in the resin formulation and pre-impregnating process to produce prepregs.

Prepreg Material Considerations

Prepregs require different handling procedures than wet lay-up materials. Because prepregs are made with resins that are already mixed, they must be stored in cold lockers (< 40°F/4.4°C) or in frozen storage, (≤ 0°F/-17.8°C), to slow the reaction of the resin. (*See* **Prepreg Storage and Handling**, Page 29.) Some prepreg resins have been developed that can be stored at room temperature for up to one year. These unique formulations require an elevated temperature cure. Common cure temperatures for thermoset-matrix prepregs range from room temperature to > 600°F, depending on the matrix.

Prepreg Manufacturing Methods

Several common manufacturing methods are used to impregnate the resin into the fiber form. In the "resin-bath" or "solvent bath" system, solvent is added to the mixed resin to dilute it and to permit rapid saturation of the dry fiber form. This "A-staged" prepreg is then taken through a heating tower to evaporate most of the solvent, raising the viscosity of the resin, and creating a "B-staged" prepreg. The resin bath method is most applicable to woven forms as it is difficult to do with uni-tape or stitched forms.

[handwritten: = B-staged P. P]

Another method is called "hot-melt" coating. There are two methods of hot-melt coating used in the prepreg industry; 1) hot-melt solution route and 2) hot-melt film route. In these techniques, resin is not diluted with solvent, but instead is quickly heated to reduce the viscosity to a suitable value for

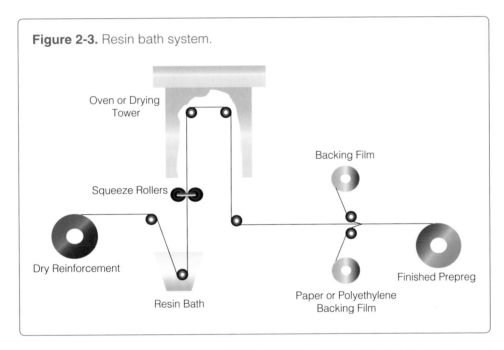

Figure 2-3. Resin bath system.

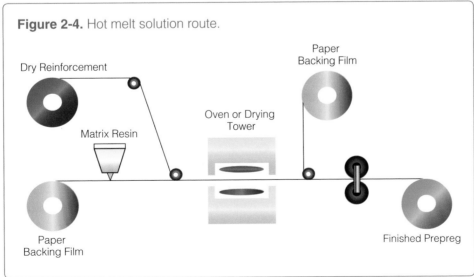

Figure 2-4. Hot melt solution route.

quick distribution into a film and/or saturation into a dry cloth. The resin may be chilled to raise the viscosity and minimize cross-linking prior to rolling it up on the roll. Often the majority of resin in hot-melt systems is coated on one side of the cloth or unidirectional tape. Hot-melt systems typically have a low volatile content as a result of this process.

Stages of a Resin System

For thermosetting resins, there is an initial A-stage, a secondary B-stage, and a final C-stage, which describes its status regarding chemical reaction and cure.

A-Stage

A-stage describes the first reaction stage for thermoset resins, characterized by low viscosity and the ability to be diluted with solvents. Prepreg manufacturers use resins in this stage to easily impregnate cloth.

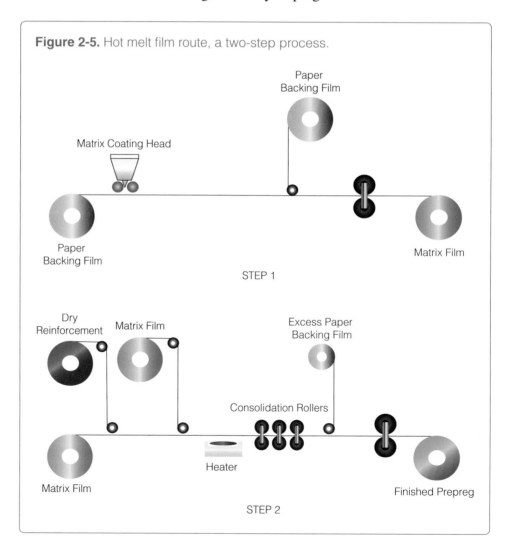

Figure 2-5. Hot melt film route, a two-step process.

B-Stage

Prepreg resins are advanced to this intermediate B-stage, where the resin is somewhat viscous and tacky, but not flowing. This facilitates storage, handling, and layup of the material. When heat is later applied during the cure process, the resin viscosity will drop and flow and the chemical reaction (cross-linking) will proceed.

C-Stage

This is the final reaction stage for thermoset resins, in which the resin has cross-linked to the point where it is insoluble and infusible. Usually prepreg in this stage is not used to make parts, due to very poor handling characteristics and bonding between plies.

Prepreg Storage and Handling

Tack and Drape

Tack refers to the stickiness of prepreg at room temperature. Prepreg manufactures may have several different tack levels available of a specified resin.

Drape describes the ability of materials to conform to compound curves without wrinkling or buckling. A very drapeable prepreg is required for structures with intricate geometries, like ducting. Drape varies greatly, dependent on both fiber/fabric form and the viscosity of the matrix system.

Prepreg Shelf Life

Shelf life is defined as the time the materials can be stored and still be processed to the desired condition. These materials almost always require frozen storage ($< 0°F/-18°C$).

Usually, the shelf life of a frozen prepreg is between six months to one year. Some systems will allow up to one year of storage at room temperature ($< 77°F/25°C$).

Prepreg Out-Time Limits

Out-time for prepreg materials is defined as the amount of time out of frozen storage ($> 0°F/-18°C$), with specific upper temperature limits ($< 77°F/25°C$), during which prepreg can still be used to make good parts. Usually, the allowable total out time is in the range of $2 - 30$ days, sometimes longer depending on the chemistry. (*See* Table 2.5 on the next page.)

TABLE 2.5 Out-Time Tracking Card

Out-Time Record Sheet								
Batch No.	Roll No.	Time Out	Date	Freezer Temp	Time In	Room Temp	Date	Total Out-Time

Moisture Contamination

The risk of excessive moisture uptake is a big problem with frozen prepregs. This can happen by opening the frozen prepreg bag and allowing condensation to form on the exposed material. For this reason, prepreg materials should always be allowed to fully thaw to room temperature prior to removal from the moisture barrier.

With epoxy prepregs, excessive moisture can greatly affect the crosslink chemistry and weaken the matrix resin's structure. Minor moisture in an epoxy matrix may be effectively removed with an extra long vacuum debulk step prior to processing.

With cyanate ester resins, moisture can accelerate the resin and the resultant carbamate reaction creates CO_2 as a by-product. The CO_2 acts as a blowing agent and will embrittle the matrix. The presence of moisture in cyanate ester resin can also cause transesterification of the matrix (i.e., the matrix polymerizes into a linear-chained thermoplastic, instead of a cross-linked thermoset structure).

Kit-Cutting Prepreg Materials

"Kitting" new rolls by re-packaging into smaller packages or kits is recommended to preserve the material's out-time. A new roll may be cut into smaller kits and then each kit can be properly identified and placed back into frozen storage until needed. The benefit here is that the smaller kits can be thawed in a shorter period of time and the rest of the roll is not working against additional out-time. This technique is especially advantages when small quantities of material are used periodically such as in a repair station or small-scale fabrication environment.

Recertification

Prepreg materials that have exceeded their shelf life or out-time can usually be tested and re-qualified for a shorter period of time. Typically this requires that the material be tested in the lab for flow and gel properties and compared to the original certification data. This is usually done in accordance with specific ASTM (American Society for Testing and Materials) standards.

Curing Thermoset Prepregs

Prepreg materials often require an elevated temperature process to facilitate the cure reaction. This is usually done under pressure in a heated press or within a vacuum bag in the oven or autoclave. Historically, a time/temperature recipe was prescribed with specific heating rates and **isothermal** holds to produce the desired state of cure (legacy cure cycle).

Processing prepreg parts per these recipes can be tricky. This is especially true when the measured process conditions go outside of the specified boundaries of the recipe, bringing to question the true cure-state of the material. Often, when such "out-of-spec" conditions happen the prepreg parts may be rejected, scrapped, or otherwise subjected to further testing because of the unknown properties of the material.

To better understand the nature of how a thermosetting prepreg material cures, we need to examine the change in viscosity and modulus of the resin as it cures during the elevated temperature cycle. The term "viscoelastic" is used in the following text to describe both the viscous (liquid) initial phase of the resin and the final elastic (hardening) stage of the resin/laminate during completion of the cure reaction.

A detailed example of how the viscoelastic properties of a thermoset resin change during a thermal cycle are outlined in Figure 2-6 (on the next page) and in the text that follows. The annotations on the graph in Figure 2-6 point to significant changes in the resin during the cure of a typical epoxy/carbon fiber prepreg:

- The viscosity drops as a function of temperature. This is measured with a torsion-plate rheometer. (Reference blue line, "APA1 Visc p")
- The chemical reaction is measured by the heat that is released during the cure. The maximum heat evolution corresponds to the rapid rise in viscosity and modulus. This is measured via differential scanning calorimeter (DSC). (Ref. green line, "DSC Heat Flow")
- Both the viscosity (flowing liquid) and rise in modulus (springiness of the solid, reference red line, "APA1 Gp") exhibit a similar change at the beginning of the cure cycle. Because the sample is a prepreg, both fibers and resin are present and as the resin viscosity rises and falls it causes more or less bending of the fibers and therefore the elastic response.
- Once the resin solidifies the flow of the resin becomes immeasurable, the apparent viscosity (blue line) will drop on the graph. This is due to a lack of molecular friction in the solid and not an actual drop in viscosity.

Figure 2-6. Example of viscosity and modulus vs. time/temperature.

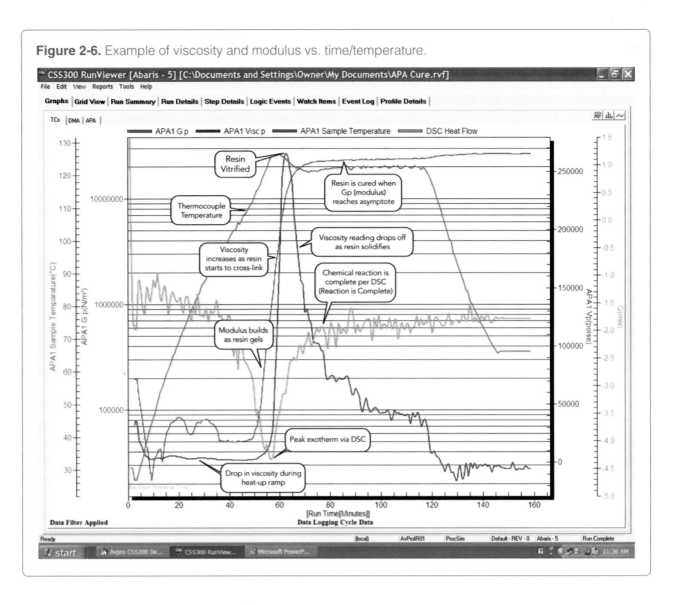

Flow and Gel

The definition of the "flow" phase is when the resin is at its lowest viscosity and maintains this viscosity for a period of time. The measured drop in viscosity and the time that the resin maintains this low-level of viscosity is highly dependent upon the temperature ramp-rate. (*See* **Too-Fast/Too-Slow Ramp Rate Problems,** Page 34.)

The interpretation of "gel" varies since it is actually the transition phase from a liquid to a solid state. Prepreg resins start out as a true liquid and end as a true solid. The gel-point is technically measured at the point where the modulus is equal to the viscosity. (Where the red and blue lines cross on the graph as the viscosity and modulus build.) However, the most easily observed, and typically referenced gel-point is at the point where the viscosity is at its peak.

Optimal Compaction Point

During the later part of the flow phase, where the resin viscosity is starting to rise, is the optimal time to add pressure and compact a laminate. (When the resin viscosity is about equal to the original B-stage viscosity of the prepreg when it was at room temperature.)

Compacting at this point ensures that all of the air and gas has had sufficient time to escape and that the resin will not re-soften or be "blown out" of the laminate with expanding gases. Lower void content can be achieved in laminates processed using this "optimized compaction" method.

Unfortunately, it is a common practice to apply pressure at the beginning of the cure cycle because of a lack of actual resin flow/gel information on all of the parts being processed in a production load. Under these conditions it is critical to remove as much of the trapped gasses as possible prior to starting the cure cycle. This can primarily be achieved by placing the laminate under vacuum for a long period of time before processing (vacuum debulking).

Vitrification and Cure

Eventually, vitrification takes place (the resin becomes a glassy solid) at a given temperature. If the temperature is raised further before full cure time is achieved, this glassy solid will re-soften. When the T_g is above a specified temperature value, the resin is considered to be "cured" and is structurally functional below this temperature.

Cure Cycles

Different "cure cycles" are specified for different prepreg resins, depending on their chemistries. For example, epoxy resins may be processed with a single or dual (heat) ramp/soak cycle with simple vacuum and pressure steps to get the best results. Bismaleimide and Polyimide resins, on the other hand, may require a multi-step ramp/soak, high temperature cycle, with several different intermittent pressurization steps to do the job.

Most legacy cure cycles are simply recipes that define time/temperature and vacuum and pressure requirements at specified time intervals. Originally these cure cycles were derived from the viscoelastic properties of the different resins as measured in the laboratory and then mapped-out as a standard time/temperature recipe with upper and lower time and temperature limitations.

One can see how the basic recipes perform on carbon epoxy prepreg samples when the viscoelastic properties are tested and analyzed in the laboratory per the prescribed heat rate and soak time at temperature in Figures 2-7 and 2-8 below (Pages 35 and 36).

Predictable Viscoelastic Change

As long as the laminate sample is actually heated at the prescribed rate and meets the target temperature for the specified time, the viscoelastic properties change as predicted. In fact, the graphs indicate that the modulus (*See* Figures 2-7 and 2-8 screenshots, "APA1 Gp," blue line) becomes **asymptotic** about midway through the soak in the single ramp profile and even earlier in the two-step process, indicating that the resin is probably fully cured at that point.

The information gleaned from these graphs seem to indicate that the soak time is longer than necessary, and in the later case, much longer than needed to cure this resin. This extra time at temperature may or may not be necessary to achieve full properties but since it is often specified that the time/temperature requirements be documented in accordance with the legacy recipe, it has become common practice to run material for the entire duration to "ensure" a full cure.

Practical Considerations

In theory this recipe should work for this resin within the target parameters given, and with a bit of a safety factor added to accommodate slight variations in the process (including shelf life and out-time factors). In practice it is often more difficult to meet the prescribed ramp rate and soak time/temperature requirements for every part in a large load of parts in an oven or autoclave, as there can be many factors affecting the ability to heat everything uniformly. (*See* Chapter 7, **Introduction to Tooling**, for more information.)

Because of these variables, actual thermocouple measurements in large size loads of real prepreg parts often demonstrate a high degree of variability and many can be out of the boundaries of the recipe or specification. This often results in parts that see too slow or too fast of a heat rate, or too long at the soak temperature in some parts to accommodate the lagging thermocouple readings in other parts.

This makes it difficult to run large loads of various sized parts and be sure that all of the parts have fully cured and/or are not held for unnecessarily long periods of time per the legacy cure cycles. Future process specifications most likely will include options to the recipe-type of cure cycles that use actual or statistically-predicted viscoelastic properties based on actual thermocouple feedback on multiple parts using new generation computer process control capabilities. (*See* **Computer Controlled Processing** below, Page 38.)

Too-Fast/Too-Slow Ramp Rate Problems

The issue of a *too-fast* or *too-slow* heat ramp can be somewhat confusing. Unfortunately, the complex relationship among viscosity, pressure, cure rate, gas devolution, etc. is a fact of life that must be managed. If the resin is heated too quickly, the flow time is shortened and the viscosity will not stay low for a long enough period of time. (*See* Figure 2-9 on the next full-page spread.)

This can diminish the exchange of volatiles out of the laminate and interrupt the movement of resin into low-pressure sites. Rapid heating will generally

Figure 2-7. Single-step 350°F (177°C) legacy cure profile.

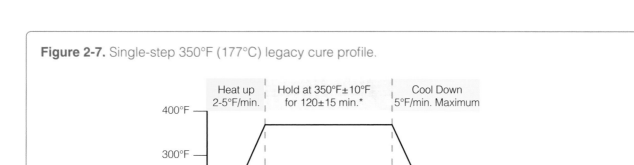

Heat up 2-5°F/min.

Hold at 350°F±10°F for 120±15 min.*

Cool Down 5°F/min. Maximum

Below 150°F release pressure and add vacuum.*

Apply 85 psi (+15/-0) pressure for laminate. Vent vacuum bag to atmosphere when pressure reaches 20 psi.

Apply 24 in. Hg/in. vacuum minimum.

*Temp. based on lagging thermocouple.

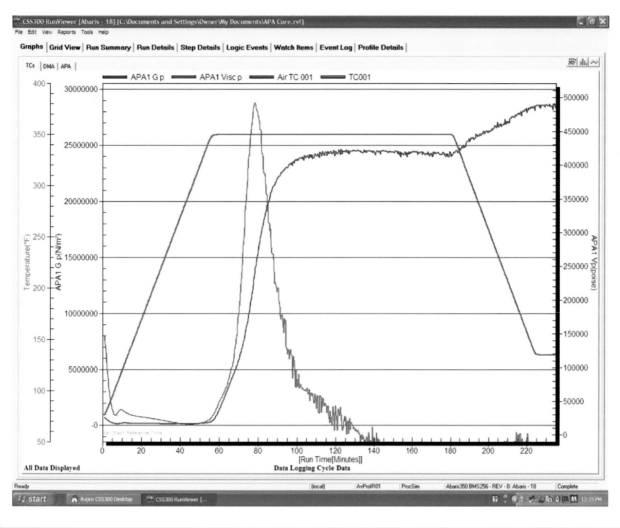

Figure 2-8. Two-step 350°F (177°C) legacy cure profile.

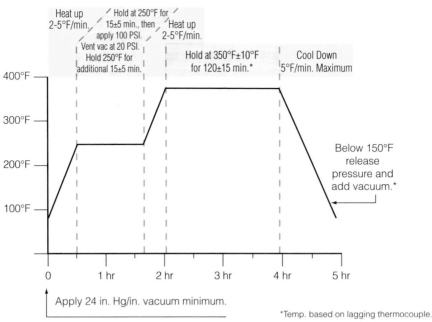

Heat up 2-5°F/min.

Hold at 250°F for 15±5 min., then apply 100 PSI.

Vent vac at 20 PSI. Hold 250°F for additional 15±5 min.

Heat up 2-5°F/min.

Hold at 350°F±10°F for 120±15 min.*

Cool Down 5°F/min. Maximum

Below 150°F release pressure and add vacuum.*

Apply 24 in. Hg/in. vacuum minimum.

*Temp. based on lagging thermocouple.

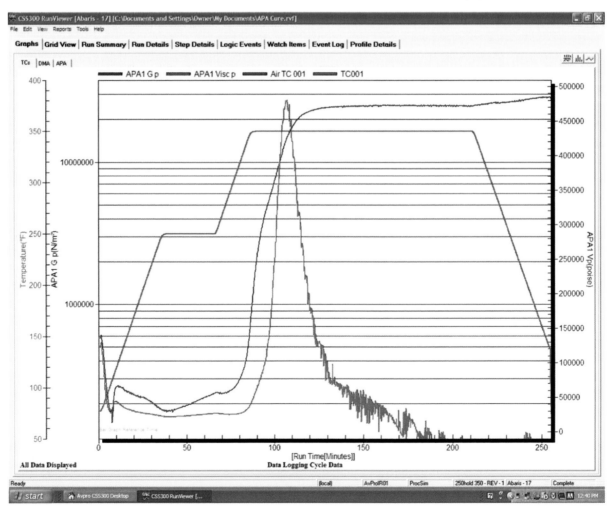

cause the resin viscosity to drop to a lower value than slow heating. If air or moisture is present, a rapid rise in temperature will cause the gas to rapidly expand and the trapped moisture to boil, thus creating resin voids and potentially displacing resin out of the laminate. In addition, too much heat, too fast, can generate an uncontrolled exothermic reaction in thicker structures that can greatly affect the resin properties.

If the prepreg is heated too slowly (*see* Figure 2-10 on the next page), the viscosity may never get low enough to allow sufficient resin movement to take place. As a consequence, trapped air and gas in the laminate may not adequately move through the viscous liquid and may result in a higher than normal void content in the laminate.

This can be a problem with a honeycomb sandwich structure as the heat-rate directly affects the flow of the resin/adhesive to the edge of the core cells. If good filleting of these cells is not achieved, a poor bond is the result. (Slow heating molds made of high-mass materials can contribute to this effect.)

Figure 2-9. "Too-fast" ramp rate.

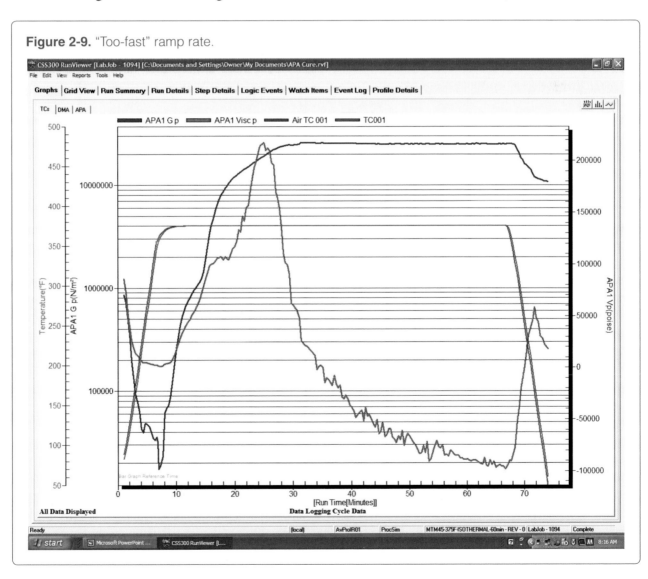

Figure 2-10. "Too-slow" ramp rate.

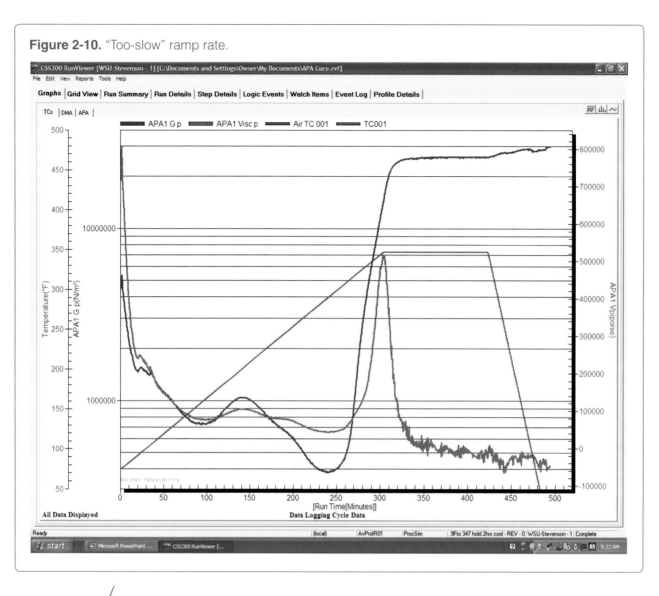

Too Long at Soak Problems

Many prepreg thermoset resin systems are cured at temperatures that push or exceed their own service temperature limits. Long soaks at these upper temperatures may actually begin to degrade, rather than improve, the matrix properties. For this reason an upper soak-time tolerance is usually specified in the cure cycle. Because of this issue, care must be taken not to exceed the maximum time boundary just to facilitate lagging thermocouple readings in the process.

Computer Controlled Processing

New computer-aided processing technology allows us to monitor and control to actual material state conditions, optimizing the thermal-process cycle, and ensuring that a full cure is achieved in the minimum amount of time possible. Vacuum and pressure targets during the process can be adjusted by the computer based on actual viscosity changes in the curing prepreg when

Figure 2-11. Isothermal cycles at different temperatures vs. time.

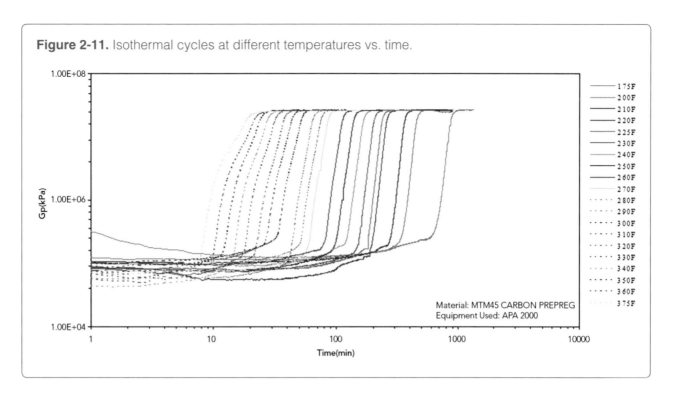

coupled with real-time analytical equipment (i.e.; DSC and/or torsion-plate rheometer).

In order to facilitate such capabilities, specifications can be written that allow options to time/temperature legacy profiles for curing prepreg composite parts. Such specifications model the viscosity and modulus change of the material and the completion of the reaction instead of relying on the legacy recipe cure cycle approach.

The value of processing to **material state conditions** using computer process controls is found in the inherent quality assurance of such a system. Not only does this allow for a more accurately controlled cure of the composite materials, but also provides a myriad of sensor feedback on large production loads and the data storage capabilities relevant to each cycle. This data can be used by both quality assurance and materials engineering for a multitude of purposes from quality control and statistical analysis to lean production improvement in processing.

Another advantage of processing with material state feedback is the ability to adjust the time/temperature cycle as needed to build large parts with minimal effect from thermal expansion of the tooling. Figure 2-11 depicts a range of time/temperature options for a selected prepreg material using the storage modulus (Gp) feedback as an indication of when the T_g exceeds the cure temperature. The resin will remain solid as long as the T_g is not exceeded at any time during an elevated temperature post cure or in-service.

Fiber Reinforcements

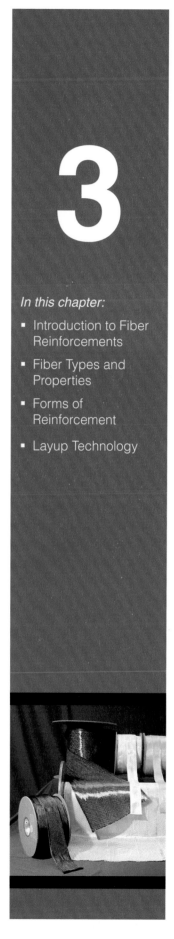

3

In this chapter:
- Introduction to Fiber Reinforcements
- Fiber Types and Properties
- Forms of Reinforcement
- Layup Technology

Introduction to Fiber Reinforcements

This section contains information about fiber types, fiber properties, and the different fiber forms that are used in the fabrication of advanced composite structures. Table 3.1 on the next page lists the most common fiber reinforcements used in manufacturing advanced composites, as well as their structural and thermal properties.

In this section we compare the properties of a variety of common fibers to those of basic glass fiber (E-glass). Everything from the weight of the fiber, to the strength and stiffness, to the decomposition temperatures are examined.

Understanding these fundamental fiber properties is the key to understanding why certain fibers are chosen for a given application. For example, glass fiber might be chosen instead of carbon for reinforcement in a composite fairing panel because it is high strength and durable and a bit more flexible than one made with carbon fiber. On the other hand, carbon fiber reinforcements might be the choice for a highly loaded spar or floor beam application where the high stiffness to weight properties would be an obvious advantage in the design.

In the segment on **Forms of Reinforcement**, we look at everything from the smallest filaments to strands, tows, yarns, mats, fabrics, braids and other forms.

Finally, basic terms used to describe reinforcements are defined as they are used in these disciplines in the **Layup Technology** and **Ply Kitting and Orientation Terminology** sections of this chapter.

41

TABLE 3.1 Common Composite Reinforcement Fibers and Properties

Fiber	Type	Density oz/in³ (g/cm³)	Tensile Strength Ksi (GPa)	Tensile Modulus Msi (GPa)	Elongation to break %	Decomposition Temp.
Glass	E-glass	1.47 (2.54)	450–551 (3.0-3.8)	11.0–11.7 (75.8-80.7)	4.5–4.9	1346°F 730°C
	S-2 glass	1.43 (2.48)	635–666 (4.3-4.6)	12.8–13.2 (88.3-91.0)	5.4–5.8	1562°F 850°C
Carbon	PAN-SM*	1.04 (1.8)	500–700 (3.4-4.8)	32–35 (221-241)	1.5–2.2	6332°F 3500°C
	PAN-IM*	1.04 (1.8)	600–900 (4.1-6.2)	42–43 (290-297)	1.3–2.0	
	PAN-HM*	1.09 (1.9)	600–800 (4.1-5.5)	50–65 (345-448)	0.7–1.0	
	Pitch-LM*	1.09 (1.9)	200–450 (1.4-3.1)	25–35 (172-241)	0.9	6332°F 3500°C
	Pitch-HM*	1.2 (2.0)	275–400 (1.9-2.8)	55–90 (379-621)	0.5	
	Pitch-UHM*	1.24 (2.15)	350 (2.4)	100–140 (690-965)	0.3–0.4	
Aramid	Twaron® 1000	0.84 (1.45)	450 (3.1)	17.6 (121.3)	3.7	842°F 450°C
	Kevlar® 29	0.83 (1.44)	525 (3.6)	12.0 (82.7)	4.0	
	Kevlar® 49	0.83 (1.44)	525 (3.6)	18.9 (130.3)	2.8	
	Kevlar® 149	0.85 (1.47)	508 (3.5)	26.9 (185.5)	2.0	
PE (Polyethylene)	Spectra® 900	0.56 (0.97)	380 (2.6)	11.5 (79.3)	3.6	302°F 150°C
	1000		447 (3.1)	14.6 (100.7)	3.3	
	2000		485 (3.3)	18.0 (124.1)	2.8	
PBO (Polyphenylene benzobisoxazole)	Zylon® AS	0.89 (1.54)	841 (5.8)	26.1 (180.0)	3.5	1202°F 650°C
	Zylon® HM	0.90 (1.56)	841 (5.8)	39.2 (270.3)	2.5	
Other Ceramic Fibers	Boron	1.49 (2.58)	522 (3.6)	58 (400)	--	3769°F 2076°C
	Silicon Carbide (SiC)	1.45–1.73 (2.5-3.0)	360–900 (2.5-6.2)	27–62 (186-428)	0.7–2.2	4172°F 2300°C
	Alumina-silica	1.76 (3.05)	250–290 (1.7-2.0)	22–28 (152-193)	0.8–2.0	1400°F 760°C
	α-alumina	1.97–2.37 (3.4-4.1)	260–510 (1.8-3.5)	38–67 (262-462)	--	1800°F 982°C
	Quartz	1.25 (2.17)	493 (3.4)	10.0 (68.9)	5.0	1920°F 1050°C

*Carbon fiber is classified by the stiffness or modulus designation:

- SM = Standard Modulus
- LM = Low Modulus
- IM = Intermediate Modulus
- HM = High Modulus
- UHM = Ultra High Modulus fiber

Fiber Types and Properties

Glass Fiber

Note: Text from this section is adapted from "The Making of Glass Fiber" by Ginger Gardiner, *Composites Technology Magazine*, April 2009; reproduced by permission of *Composites Technology Magazine*, copyright 2009, Gardner Publications, Inc., Cincinnati, Ohio, USA.

Fiberglass yarns are available in different formulations. "E" glass (electrical) is the most common all-purpose glass fiber, while "S-2" (high strength) glass is used for special applications.

Manufacture

Textile-grade glass fibers are made from silica (SiO_2) sand which melts at 1,720°C/3,128°F. Though made from the same basic element as quartz, glass is amorphous (random atomic structure) and contains 80 percent or less SiO_2, while quartz is crystalline (rigid, highly-ordered atomic structure) and is 99 percent or more SiO_2. SiO_2 is molten at roughly 1,700°C/3,092°F, and if cooled quickly will not form an ordered, crystalline structure but will instead remain amorphous—i.e., glass. Although a viable commercial glass fiber can be made from silica alone, other ingredients are added to reduce the working temperature and impart other properties useful in specific applications. (See Table 3.2, "Comparison of E and S-2 glass fibers.")

TABLE 3.2 Comparison of E and S-2 glass fibers

Composition	E-glass	S-2 Glass®*
Silicon Dioxide	52–56%	64–66%
Calcium Oxide	16–25%	
Aluminum Oxide	12–16%	24–26%
Boron Oxide	8–13%	
Sodium & Potassium Oxide	0–1%	
Magnesium Oxide	0–6%	9–11%

*S-2 GLASS® is a registered trademark of AGY.

For example, E-glass, originally aimed at electrical applications, with a composition including SiO_2, Al_2O_3 (aluminum oxide or alumina), CaO (calcium oxide or lime), and MgO (magnesium oxide or magnesia), was developed as a more alkali-resistant alternative to the original soda lime glass. S-glass fibers (i.e., "S" for high strength) contain higher percentages of SiO_2 for applications in which tensile strength is the most important property.

Glass fiber manufacturing begins by carefully weighing exact quantities and thoroughly mixing (batching) the component ingredients. The batch is then melted in a high temperature (~1,400°C/2,552°F) natural gas-fired furnace.

Beneath the furnace, a series of four to seven bushings are used to extrude the molten glass into fibers. Each bushing contains from 200 to as many as 8,000 very fine orifices. As the extruded streams of molten glass emerge from the bushing orifices, a high-speed winder catches them and, because it revolves very fast (~2 miles/3 km per minute—which is much faster than the speed the molten glass exits the bushings), tension is applied and this draws the glass streams into thin filaments (i.e., fibrous elements ranging from 4–34 µm in diameter, or 1/10 that of a human hair).

Fiber Diameter and Yield

The bushings' orifice or nozzle diameter determines the diameter of the glass filament; nozzle quantity equals the number of ends. A 4,000-nozzle bushing may be used to produce a single roving product with 4,000 ends, or the process can be configured to make four rovings with 1,000 ends each. The bushing also controls the fiber yield or yards of fiber per pound of glass. (The metric unit, tex, measures fiber linear density: 1 tex = 1 g/km; while yield is the inverse, yd/lb.) A fiber with a yield of 1,800 yd/lb (275 tex) would have a smaller diameter than a 56 yd/lb (8,890 tex) fiber, and an 800-nozzle bushing produces a smaller yield than a 4,000-nozzle bushing. For example, OCV Reinforcements (Toledo, Ohio) commonly uses a 4,000-nozzle bushing for optimizing production flexibility, while AGY uses 800-orifice bushings because it is a smaller company that produces finer filaments and smaller-run niche products. The range of glass fiber diameter, or micronage, has become more varied as composite reinforcement applications have become more specialized. OCV observes 17 µm and 24 µm as the most popular glass fiber diameters for composite reinforcements, although its reinforcement products vary from 4 µm to 32 µm. AGY's products typically fall into the 4 µm to 9 µm range.

Size vs. Finish

In the final stage of glass fiber manufacturing, a chemical coating, or **size**, is applied. (Although the terms binder, size, and sizing often are used interchangeably in the industry, size is the correct term for the coating applied and sizing is the process used to apply it.) Size is typically added at 0.5 to 2.0 percent by weight and may include lubricants, binders, and/or coupling agents. The lubricants help to protect the filaments from abrading and breaking as they are collected and wound into forming packages and, later, when they are processed by weavers or other converters into fabrics or other reinforcement formats.

Coupling agents cause the fiber to have an affinity for a particular resin chemistry, improving resin wet-out and strengthening the adhesive bond at the fiber-matrix interface. Some size chemistry is compatible only with polyester resin and some only with epoxy, while others may be used with a variety of resins.

AGY, OCV and PPG agree that size chemistry is crucial to glass fiber performance and each company considers its size chemistry proprietary. PPG believes that in many composite applications, performance can be achieved via size chemistry as effectively, if not more so than with glass batch chemistry. For example, its 2026 size chemistry used with HYBON products for wind blades reportedly achieves an order of magnitude improvement in blade fatigue life by significantly improving the fiber's ability to be wet-out by and adhere to all resin types.

After textile glass fibers are processed by weaving, the completed fabric is heat cleaned to remove any remaining size, and then a finish is applied. Finishes are applied to glass fabrics to improve resin wet-out and enable a strong bond between the resin and the glass fibers. Some finishes are just for polyester, some are just for epoxy, and some are good for many resins. An incorrectly selected finish can greatly reduce the ability of the matrix to wet and bond to the fiber, which subsequently diminishes the overall laminate properties.

Volan (chromium methacrylate) and Silane (silicon tetrahydride) are common finishes. Volan is much older and was originally developed to promote adhesion to polyester resins, although it will bond to epoxies. Silane is more common for composites used in aircraft because it forms a stronger bond with epoxy resins. Volan has traditionally been a softer finish giving a more pliable fabric, but recent developments have produced some excellent soft-silane finishes.

Glass Fiber Products History and Development

E-glass, ECR-glass and ADVANTEX

The industry norm for composite reinforcement worldwide is E-glass; S-glass is the standard for higher-performance applications such as aerospace and ballistics. A major evolution for E-glass has been the removal of boron. Although at one time boron facilitated fiberization, it is expensive and produces undesirable emissions. Its removal has produced multiple benefits including lower cost and a more environmentally-friendly glass fiber.

OCV's boron-free product, ADVANTEX, is actually its second-generation ECR-glass. Its first iteration in the 1980s was a response to a market need for even higher corrosion resistance coupled with good electrical performance. However, this original patented ECR-glass was difficult to make and therefore more expensive to end-users. Thus, OCV developed ADVANTEX, which is more cost-effective due to a lower-cost, boron-free batch composition and the elimination of scrubbers and other environmental equipment previously required to capture boron emissions. ADVANTEX is also made using higher temperatures, which produces higher properties.

OCV is converting all of its global reinforcements manufacturing to ADVANTEX, including the 19 Saint-Gobain Vetrotex reinforcements plants it acquired in 2007. Its patent on ECR-glass has recently expired, enabling companies like Fiberex (Leduc, Alberta, Canada) and CPIC (Chongqing, China) to emerge with their own versions.

S-2 and S-3 GLASS, T-glass and HS2/HS4

S-glass has also evolved. Driven by U.S. military missile development in the 1960s and its need for high-strength, lightweight glass fiber for rocket motor casings, Owens Corning pioneered the production of S-glass, and subsequently developed an improved form trademarked as **S-2 GLASS**, featuring a 40% tensile strength and 20% higher tensile modulus than E-glass. These properties are derived from its composition, though the manufacturing process helps to maintain that performance, as does using the correct size for the polymer matrix in the final composite structure.

This business, Owens Corning's fine glass yarns and S-2 GLASS fiber products, was spun off in 1998, then reorganized and emerged in 2004 as AGY. S-3 GLASS is described as "designer glass," in that AGY can make small amounts (e.g.,100 tons) of it customized to meet one customer's precise specifications. One example is AGY's HPB biocompatible glass fiber, developed for long-term medical implants (over 30 days) such as orthodontics and orthopedics.

Other higher-performance products include T-glass manufactured by Nittobo (Tokyo, Japan) and Sinoma Science & Technology's (Nanjing, China) HS2 and HS4 products, distributed in Europe and North America exclusively by PPG.

S-1 and R-glass

Higher-performance glass fiber types have traditionally been harder to produce, requiring higher melt temperatures and paramelt processing using smaller paramelter furnaces with low throughput-all of which increases cost. Both AGY and OCV saw a need for higher performance glass fiber at a lower cost. AGY developed S-1 GLASS, situated between E-glass and S-2 GLASS in performance and cost, and targeted for composite wind blades where its higher properties reduce the amount of glass fiber required as blade lengths are extended.

OCV developed its similarly situated high-performance glass (HPG) process in 2006, which produces higher-performance glass fibers on a larger scale. This process uses furnaces smaller than those used to make all-purpose E-glass, but 50 times larger than a paramelter, resulting in lower cost than S-glass. The resulting OCV products are all R-glass by composition—FliteStrand, WindStrand, XStrand and ShieldStrand—and feature different fiber diameters (12–24 µm), mechanical properties, and size chemistry, specific to the performance required for each end-use. For example, ShieldStrand requires a lower micronage of 12.5 µm, while WindStrand, aimed at very large wind blade structures, requires a greater micronage of 17 µm. WindStrand features a tensile modulus not quite as high as S-glass, but higher than E-glass at an affordable cost. The size applied to ShieldStrand enables progressive delamination of the composite armor. It does this by allowing separation at the fiber-matrix interface upon impact, yet maintaining static mechanical properties of the composite.

TABLE 3.3 Comparison of Mechanical Properties for Various Glass Fiber Products

	E-glass	R-glass	HS2/HS4	T-glass	S-2 GLASS
Tensile Strength (GPa)	1.9 – 2.5	3.1 – 3.4	3.1 – 4.0	not available	4.3 – 4.6
Tensile Modulus (GPa)	69 – 80	89 – 91	82 – 90	84	88 – 91

TABLE 3.4 Types of Glass Fiber

A	High-Alkali
E	Electrical
S	Structural: ½ fiber diameter of E-glass; stronger
S-2*	Commercial grade S-glass, has larger filaments
Advantex®	Produced as a replacement for E-glass and E-CR Glass
D	Dielectric grade
M	High-modulus (boron-rich) glass
C	Chemical resistance
E-CR	High chemical resistance
R	High strength-high modulus
Lead	Radioactivity resistance
Lithium-oxide	X-Ray transparency

*S-2 GLASS® is a registered trademark of AGY.

Typical Glass Fiber/Fabric Forms and Styles

(This information is provided by Hexcel-Schwebel, *Technical Fabrics Handbook*.)

The wide variety of fiberglass yarns produced requires a special system of nomenclature for identification. This nomenclature consists of two parts: one alphabetical and one numerical.

Example: ECG 150-1/2 (U.S. System)

First Letter (E)—Characterizes the glass composition (E-glass)

Second Letter (C)—Indicates the yarn is composed of continuous filaments. S indicates staple filament. T indicates texturized continuous filaments.

Third Letter—Denotes the individual filament diameter, in this case the G indicates the diameter is 0.00036 inches (9 microns) in diameter. (Commercial glass filaments range in size from 0.00017 – .00051 inches (4 – 13 microns) in diameter).

First Number (150)—Represents 1/100 the normal bare glass yardage in one pound of the basic strand. In the above example, multiply 150 by 100, which results in 15,000 yards in one pound.

Second Number (1/2)—Represents the number of basic strands in the yarn. The first digit represents the original number of twisted strands. The second digit separated by the diagonal represents the number of strands plied (or twisted) together. To find the total number of strands used in a yarn, multiply the first digit by the second digit (a zero is always multiplied as 1).

Note that literally hundreds of styles are available from various manufacturers; these examples are only a few of the many available. Unidirectional forms are also available. (*See* Table 3.5.)

TABLE 3.5 Common Fiberglass Fabric Weave Styles

Hexcel-Schwebel Style Number	Weave Style	Yarn Count Yarns per inch Warp/Fill	Fabric Weight oz/sq. yd (g/m^2)	Fabric Thickness inches (mm)	Common U.S. Fiber Nomenclature
120	4 HS	60/58	3.16 (107)	0.0035 (0.09)	ECG 450 1/2
1581	8 HS	57/54	8.90 (301)	0.008 (0.20)	ECG 150 1/2
6781	8 HS	57/54	8.90 (301)	0.009 (0.23)	S-2CG 75 1/0
7500	Plain	16/14	9.60 (325)	0.011 (0.28)	ECG 37 1/2
7544	2-End Plain	28/14	18.20 (617)	0.020 (0.51)	ECG 37 1/2 ECG 37 1/4
7597	Dbl. Satin	30/30	38.60 (1309)	0.041 (1.03)	ECG 37 1/4
7725	2x2 Twill	54/18	8.80 (298.0)	0.0093 (0.24)	ECG 75 1/0 ECH 25 1/0
7781	8 HS	57/54	8.95 (303)	0.009 (0.23)	ECDE 75 1/0

Carbon Fiber

Carbon vs. Graphite

Often called "graphite" fiber, the truth is that carbon fiber is considerably different in form than true graphite. Graphite is very soft, but brittle, like pencil lead. (*See* Figures 3-1 and 3-2 on the next full-page spread.)

Carbon fiber filament has an aligned crystalline structure making it stronger and more durable than graphite. Since carbon fiber is atomically the same as graphite and it has historically been called graphite fiber, the terminology is widely accepted as fact in the composite materials industry. The term carbon fiber is technically correct and therefore will be used throughout the following text to describe these fibers.

PAN vs. Pitch Precursors

Carbon fiber has a higher tensile modulus and compressive strength, and is lighter and stiffer than fiberglass, but is also more brittle, with lower impact resistance. The properties of carbon fiber vary, depending upon the type of fiber and **precursor** material used to make it. Carbon fiber is either made from polyacrylonitrile (PAN), an acrylic fiber material, or oil or coal pitch. PAN is often used to make Standard Modulus (SM) and Intermediate Modulus (IM) carbon fiber with notable tensile strength. Pitch is often used to make High Modulus (HM) and Ultra High Modulus (UHM) carbon fibers that are very stiff and somewhat brittle.

Many carbon fiber manufacturers designate their high and ultra high modulus pitch (or synthetic mesophase pitch) based fibers as "graphite" fibers in their product literature. These higher modulus fibers are processed at temperatures much like graphite, however in a filament form that is aligned and pulled in tension. These fibers tend to have a much higher density, a lower tensile strength, and are considerably more brittle than any PAN-based carbon fibers.

Manufacture

Carbon fiber filaments are made by **carbonizing** the precursor materials at high temperatures in an inert gas environment (*see* Figure 3-3). The carbonized filaments are pulled or aligned in tension as the material is processed to give the fiber high tensile and modulus properties.

After processing, the carbon fiber is again oxidized by immersing the fibers in air, carbon dioxide, or ozone. This can also be accomplished with an immersion bath in sodium hypochlorite or nitric acid. The fiber is then treated with the appropriate finish for bonding to the matrix resin.

Carbon Fiber/Fabric Forms and Styles

Unidirectional tape is one of the most common forms of carbon fiber reinforcement. Uni-tape is typically available in widths of 6, 12 and 24 inches, with cured ply thickness ranging from 0.003 to 0.009 inches (0.08 to 0.23 mm).

TABLE 3.6 Carbon Fiber Unidirectional Tape

Grade	Aerial Weight oz/yd^2 (g/m^2)	Thickness Inches (mm)
95	2.8 (95)	0.0045 (0.11)
145	4.3 (145)	0.006 (0.15)
180	5.3 (180)	0.0075 (0.19)
190	5.6 (190)	0.008 (0.20)

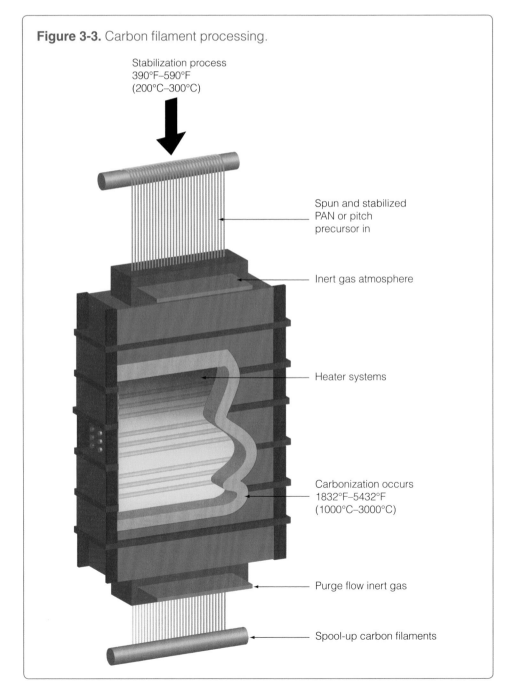

Figure 3-1. Monolithic graphite.

Figure 3-2. Carbon tow.

Figure 3-3. Carbon filament processing.

Stabilization process
390°F–590°F
(200°C–300°C)

Spun and stabilized PAN or pitch precursor in

Inert gas atmosphere

Heater systems

Carbonization occurs
1832°F–5432°F
(1000°C–3000°C)

Purge flow inert gas

Spool-up carbon filaments

Table 3.7 below shows some of the many weave patterns and fabric styles available for woven carbon fiber reinforcements.

TABLE 3.7 Carbon Fiber Fabric Weave Styles

Hexcel-Schwebel Style Number	Yarn Size	Fabric Weave Style	Yarn Count Yarns per inch Warp/Fill	Fabric Weight oz/sq. yd. (g/m^2)	Fabric Thickness Inches (mm)
130	1 K	Plain	24 /24	3.7 (125)	0.0056 (0.14)
282	3 K	Plain	12.5/12.5	5.8 (197)	0.0087 (0.22)
284	3K	2x2 Twill	12.5/12.5	5.8 (197)	0.0087 (0.22)
433	3 K	5-HS	18/18	8.4 (285)	0.0126 (0.32)
444	3 K	2x2 Twill	18/18	8.4 (285)	0.0126 (0.32)
584	3 K	8-HS	24/24	11.0 (373)	0.0165 (0.42)
613	6 K	5-HS	12/12	11.1 (376)	0.0166 (0.42)
670	12 K	2x2 Twill	10.7/10.7	19.8 (671)	0.0297 (0.75)
690	12 K	2x2 Basket	10/10	18.7 (634)	0.0280 (0.71)

Other Fibers

Quartz Fiber

Quartz fiber can be used at much higher temperatures than either E-glass or S-glass fiber, up to 1,050°C (1,922°F) and Saint Gobain's Quartzel® 4 can be used at temperatures of up to 1,200°C (2,192°F). Quartz fiber has the best dielectric constant and loss tangent factor among all mineral fibers. For this reason, along with its low density (2.2 g/cm^3), zero moisture absorption, and high mechanical properties, quartz fiber has been used frequently in high-performance aerospace and defense radomes. Quartz fiber also has almost zero coefficient of **thermal expansion** in both radial and axial directions, giving it excellent dimensional stability under thermal cycling and resistance to thermal shock.

Quartz fiber is produced from quartz crystal, which is the purest form of silica. The crystals, which are mined mainly in Brazil, are ground and purified to enhance chemical clarity. The quartz powder is then fused into silica rods, which are drawn into fibers and coated with size for protection during further processing. Quartz fiber can be supplied as single and plied yarn, roving, chopped strand, wool, low density felt, needle-punched felt, sewing thread, braid, veil, fabric and tape. Saint-Gobain offers quartz yarn and roving in a wide range of tex (linear density) and with three different size coatings. Note that with quartz fiber, the size is maintained through textile processing and is said to improve resin wet-out and fiber compatibility with the resin matrix.

Saint-Gobain produces Quartzel® fiber products, JPS Composite Materials supplies Astroquartz® products, and CNBM International Corporation is a supplier in China.

Boron Fiber

Boron fiber was the first high strength, high modulus, low density reinforcement developed for advanced composite aerospace applications and has been in commercial production for thirty-five years. Boron is an ultra-high modulus fiber, very stiff and strong, with thermal expansion characteristics similar to that of steel. Unlike carbon fiber, boron exhibits no galvanic corrosion potential with metals.

The sole manufacturer is Specialty Materials, Inc. (Lowell, Massachusetts), who uses chemical vapor deposition (CVD) to make their boron fibers. This CVD process heats a fine tungsten wire substrate in a furnace pressurized with boron trichloride gas, which decomposes and deposits boron onto the tungsten wire.

Boron is available in 3.0, 4.0, 5.6, and 8.0 mil diameter filaments. 4.0 mil diameter is the most common size available. Boron is available in unidirectional prepreg tape only. It cannot be woven into cloth since the filaments cannot be crimped without snapping. Hy-Bor®, a boron-carbon hybrid, is also available.

Use of boron fiber in composites began in the 1970s for new part production in aerospace components. Now, it has largely been replaced by high-modulus carbon fibers in those applications. However, is has become very popular for repairs to metallic structures. Boron is also very popular for sporting goods applications, where it offsets the drawbacks of high modulus carbon fiber by offering improved damage tolerance and a compressive strength that is twice its tensile strength, resulting in a superior all-around composite structure.

Synthetic Fibers

Aramid Fiber

Aramid fibers are different from carbon and glass fibers in that they are textile fibers, spun from an organic polymer versus fibers created by drawing out an inorganic material. The properties of aramid fiber vary according to type. For example, Kevlar® 29 is used for ballistic protection and has a lower modulus and greater elongation, while Kevlar® 49 is used for structures and has a higher modulus and lower elongation. Aramid fibers are 43% lighter than, and twice as strong as E-glass, and ten times stronger than aluminum on a specific tensile strength basis. Kevlar® 149 was commercialized in 1987 and has a modulus that is 40% greater than Kevlar® 49. It was developed for specific aircraft applications, but never gained widespread usage, perhaps because it is more expensive than Kevlar® 49.

Aramid fibers have excellent resistance to high and low temperatures. They excel in tensile strength but have somewhat poor laminate-compressive properties. Their main disadvantage is that they are hydroscopic (absorb water) due to their inherent aramid chemistry. Aramid fibers have much better impact resistance than carbon fiber and are often used in "crack-stop" layers between plies of carbon. Aramid fiber is cheaper than carbon fiber but more expensive than fiberglass.

Manufacturers

Kevlar® is a **para-aramid** fiber that was developed by DuPont and introduced in 1973. It is currently produced and supplied by DuPont. Technora is also a para-aramid fiber, but uses a slightly different chemistry in its manufacturing process. It was introduced in 1976 by the Japanese company Teijin, who also makes Teijinconex, a **meta-aramid** fiber which competes with DuPont Nomex®. Twaron® is a para-aramid fiber which is basically the same as Kevlar®, originally developed in the early 1970s by the Dutch company AKU, which later became Enka (Glanzstoff), and then Akzo Nobel, but since 2000 has been owned by the Teijin Group. Thus, Teijin now produces and supplies Twaron®, and the Technora® brand name is no longer supported.

Typical Fiber/Fabric Forms and Styles

Aramid tows and yarns are defined by weight using the Denier or Decitex system. Denier weight is listed in grams/9,000 meters. For example: 1,140 denier = 1,140 grams/9,000 m. The Decitex (dtex) system measures grams/10,000 m.

TABLE 3.8 Kevlar® 49 Fabric Weave Styles

Hexcel-Schwebel Style Number	Yarn Type Denier	Fabric Weave Style	Yarn Count Yarns per inch Warp/Fill	Fabric Weight oz/ yd² (g/m²)	Fabric Thickness Inches (mm)
348	380	8-HS	50/50	4.9 (166)	0.008 (0.20)
350	195	Plain	34/34	1.7 (58)	0.003 (0.08)
352	1140	Plain	17/17	5.3 (180)	0.010 (0.25)
353	1140	4-HS	17/17	5.3 (180)	0.009 (0.23)
354	1420	Plain	13/13	4.8 (163)	0.010 (0.25)
386	2130	4x4 Basket	27/22	13.6 (461)	0.025 (0.63)

TABLE 3.9 Twaron® T-1055, T-2000, T-2200, T-2040 Fabric Weave Styles

Hexcel-Schwebel Style Number	Yarn Type Denier	Fabric Weave Style	Yarn Count Yarns per inch Warp/Fill	Fabric Weight oz/ yd² (g/m²)	Fabric Thickness Inches (mm)
5348	405	8HS	50/50	4.9 (166)	0.008 (0.20)
5352	1210	Plain	17/17	5.0 (170)	0.010 (0.25)
5353	1210	4HS	17/17	5.0 (170)	0.009 (0.23)
5704	930	Plain	31/31	6.5 (198)	0.012 (0.30)
5722	1680	Plain	22/22	8.25 (280)	0.015 (0.38)

Spectra® and Dyneema®

Spectra® and Dyneema® are ultra lightweight, high-strength thermoplastic fibers made from ultra high molecular weight polyethylene (UHMWPE), also known as high modulus polyethylene (HMPE) or high performance polyethylene (HPPE). Dyneema® was developed by the Dutch company DSM in 1979, and Spectra® was developed by Honeywell International Corporation. Though the two products are chemically identical, they have slightly different production details, resulting in mechanical properties which are not identical, but very comparable. HPPE features high tensile strength, very light weight, low moisture absorption, and excellent chemical resistance. Its high specific modulus, high energy-to-break and damage tolerance, non-conductivity and good UV resistance make it a good alternative to aramid fiber. It has limited temperature service ($< 220°$ F/$104°C$) and is used in underwater, ballistic, and body armor applications. Types include Spectra® 900, Spectra® 1000, Dyneema® SK76 for ballistic armor and Dyneema® UD-HB for hard ballistic armor applications. Fabrics made with Spectra® fiber are typically used in ballistic and high impact composite applications. Table 3.10 shows common fabric styles for Spectra®. Unidirectional forms are also available.

TABLE 3.10 Spectra® Fabric Weave Styles

Yarn Type Denier	Hexcel-Schwebel Style Number	Fabric Weave Style	Fabric Weight oz/yd² (g/m²)	Fabric Thickness Inches (mm)
215	945	Plain	2.6 (88)	0.006 (0.15)
650	951	Plain	2.9 (108)	0.011 (0.28)
650	985	8-HS	5.5 (186)	0.014 (0.36)

Zylon®

Zylon® fibers consist of a rigid chain of molecules of PBO (poly p-phenylene-2, 6-benzobisoxazole). This new fiber from Toyobo has excellent tensile strength and modulus. Fabrics made from Zylon® are found in both ballistic and structural applications.

TABLE 3.11 Zylon® Fabric Weave Styles

Yarn Type Denier	Hexcel-Schwebel Style Number	Fabric Weave Style	Fabric Weight oz/yd² (g/m²)	Fabric Thickness Inches (mm)
246	550-HM	Plain	1.8 (61)	0.003 (0.08)
500	530-AS	Plain	3.8 (129)	0.008 (0.20)
500	545-AS	Plain	6.25 (212)	0.013 (0.33)
1000	1030-AS	Plain	8.0 (271)	0.016 (0.41)
1000	1035-AS	Plain	8.85 (300)	0.017 (0.43)

Forms of Reinforcement

Composites can be categorized by the form of reinforcement used. Structural composites typically use continuous fiber reinforcement. Non-structural composites may use chopped fibers. Fibers less than 0.2 inches in length are referred to as short fibers, and milled fibers are usually not considered as reinforcement, but simply fillers.

Forms Terminology

Filament

Filament describes the basic structural fibrous element. It is continuous, or at least very long compared to its average diameter, which is usually $5-10$ microns(μm). For boron, the average filament diameter is much larger, $100-140$ μm.

Strand

A strand is like a mini-yarn composed of filaments. Strands of glass can be anywhere from 51 to 1,624 filaments in size. Glass yarns are made of single or multiple strands depending on yarn size requirements.

Yarn

A yarn is a small, continuous bundle of filaments, generally fewer than 2,000. A yarn is composed of filaments in carbon and aramid fiber, and of strands in fiberglass. The filaments are lightly stranded together so they can be handled as a single unit and may be twisted to enhance bundle integrity, very important for operations like weaving fabric. In carbon fiber, common yarn sizes are from 1K to 24K. In aramid fiber, yarns can contain from 25 to 5,000 filaments.

Tow

A tow is basically the same as a yarn, except not twisted. The number of filaments comprising a carbon tow is usually 3,000, 6,000 or 12,000 and designated by using K (e.g. 3K, 6K, 12K tow). 12K tow is the cheapest, 3K tow is the most expensive. The smaller tow sizes are normally used in weaving,

Figure 3-4. Filament, strand, yarn, and tow.

Filament– the very smallest form of the fiber as extruded or spun from the raw materials.

FILAMENT

Range from 3-25 microns

The basic element of all fiber reinforcements

Strand – a bundle of (typically glass) filaments twisted into a "mini yarn" for the subsequent weaving operations. Multiple strands than can be combined to make a specific weight of yarn.

Strands to yarn – is typical to glass yarn construction. Whereas either a single strand or a twisted bundle of multiple strands make up a larger yarn for subsequent weaving operations.

Yarn – filaments are arranged into full size twisted bundles for subsequent weaving operations.

Tow – filaments are arranged into full size parallel bundles for use in making unidirectional tape with multiple tows laid side by side, or used in a direct fiber placement process or other application.

winding and braiding applications, while large tow sizes are used in unidirectional tapes and stitched multi-axial fabrics. Very thin tapes are also made from low filament-count tow for satellite applications, and there are also very large tows (50K) made for larger, industrial-type structures that require lower cost (e.g. $8/lb versus $18–20/lb for standard modulus 12K tow).

Roving

A roving is a number of strands, yarns, or tows collected into a parallel bundle with little or no twist. It is a term usually used to describe fiberglass and sometimes aramid reinforcements. Single-end rovings are smaller in diameter and are more expensive to produce due to the higher precision process required for the smaller diameter. Multi-end rovings are larger in diameter and generally less expensive, but can be more difficult to process in **filament winding** and **pultrusion** applications. Rovings are most commonly used for constructing heavy fabrics (e.g., "woven roving").

Woven Roving

Woven roving describes a heavy, coarse fabric produced by weaving continuous roving bundles, usually in a plain weave pattern. (Also applies to basket weave construction.)

Non-woven Roving

Non-woven roving is a reinforcement composed of continuous fiber strands loosely gathered together. (Also called continuous strand mat.)

Tape

The common name for "unidirectional tape"—all flat parallel tows, available in both dry and prepreg forms. It is the strongest of the common forms for prepreg.

Fabric

Fabric is a 2-D or planar textile structure that may be woven, non-woven or stitch-bonded/knitted.

Mat

Mat describes sheets of either chopped or continuous strands of fiber, held together with a small amount of adhesive binder.

Stitched

A straight-stitched or stitch-bonded material is constructed by stitching or knitting together multiple layers of unidirectional reinforcement with non-reinforcing binder fibers. Advantages include higher load-carrying properties because fibers are not crimped, faster fabric production time versus woven fabric, and reduced materials use due to optimized properties.

Non-woven Materials

Non-woven reinforcements used in composites include materials referred to as "tissue," "mat," or "veil." They can be made with long or short discontinuous fibers (i.e., chopped) or continuous fibers. They are typically used as surface ply materials in structural laminates and as reinforcements in semi-structural applications. Available in a wide range of different materials including carbon, glass, and aramid fibers, non-woven materials can also be attached to unidirectional, bidirectional, or multiaxial fabrics and tapes if desired.

Woven Fabrics

Plain Weave

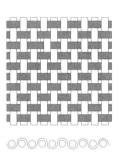

Plain weave consists of yarns interlaced in an alternating fashion, one over and one under every other yarn. The plain weave fabric provides stability, but is generally the least pliable, and least strong, due to a lower yarn count and a higher number of crimps than other weave styles. Plain weave fabrics tend to have significant open areas at the numerous yarn intersections. These intersections will be either resin rich pockets or voids in a laminate made with this fabric style. There is no warp or fill-dominant face on a plain weave fabric.

Basket Weave

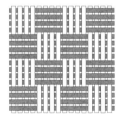

Basket weave is a plain weave except that two or more warp yarns and two or more fill yarns (rovings) are alternately interlaced over and under each other. The basket weave is also known as a "woven roving."

4-Harness Satin (4HS)

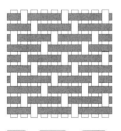

4HS weave (also known as crowfoot weave) is more pliable than the plain weave and therefore is easier to form around compound curves. In this weave pattern a fill yarn floats over three warp yarns and under every 4th yarn in the fabric. As with all harness-satin weaves, there is a dominant warp and fill face to consider in layup.

5-Harness Satin (5HS)

5-harness satin is similar to the crowfoot, except that one fill yarn floats over four warp yarns and under every 5th yarn in the fabric. The 5HS is slightly more pliable and will have a higher yarn count than the 4HS or plain weave fabrics.

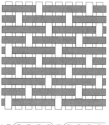

8-Harness Satin (8HS)

The 8-harness satin is similar to the 5-harness, except that one fill yarn floats over seven warp yarns and under every 8th yarn in the fabric. This is a very pliable weave, with good drape, and especially adaptable for forming over compound curved surfaces. It is typically a more expensive weave, as a higher yarn count is attainable. (8HS fabrics often have twice the yarn count of a plain weave fabric of the same yarn size).

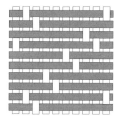

Twills

Twills are also more pliable than plain weave fabrics, allowing good conformance over compound curves. Like a plain weave, there is no warp or fill-dominant face to be concerned with in layup. The weave pattern is characterized by a visual diagonal rib caused by one fill yarn floating over at least two warp yarns, then under two, etc. (2 x 2 twill depicted). Twills have also demonstrated good impact resistance and damage tolerance.

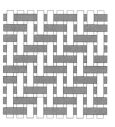

Stitched Multiaxial Fabrics

Stitched multiaxial fabrics consist of one or more layers of unidirectional reinforcement, mat, or cloth, which are stitched together. They are sometimes referred to as stitch-bonded, stitched, or knit fabrics. Stitched fabrics are not woven and therefore the yarns are not kinked or crimped like typical woven fabrics.

Stitching

The stitching thread is usually polyester due to its combination of low cost and ability to bind the layers together. However, some resins do not bond to it readily, so good resin encapsulation can be an issue.

Courses is the term for the number of stitches of yarn per inch in a stitched fabric measured in the longitudinal direction. Varying from 4 to 30, it affects the fabric's drape and wettability. *Gauge* is the number of stitches of yarn

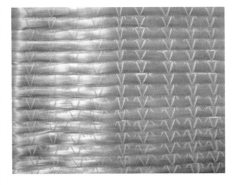

Figure 3-5. Stitched multiaxial fabric—top of fabric *(left)* and bottom *(right)*.

per inch measured in the transverse direction, and is usually a set value determined by the manufacturer.

Typically binding is achieved using a tricot stitch, which zigzags back and forth across the top of the fabric, and is actually required for any fabric using $0°$ reinforcements. A combination of stitches gives the best overall handling properties, with tricot on the top of the fabric and a chain stitch on the bottom of the fabric, running in a straight line down the warp direction, which enables drapeability.

Unidirectional

A single-layer reinforcement with all of the fibers oriented in either the longitudinal or transverse direction.

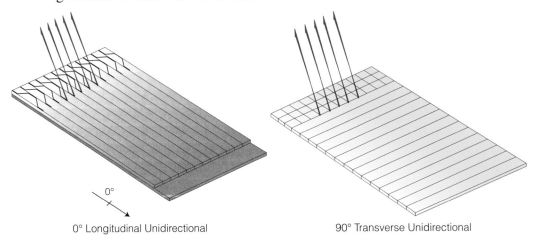

0° Longitudinal Unidirectional

90° Transverse Unidirectional

Biaxial

A two-layer reinforcement with two possible configurations. The most common is to have one layer in the $0°$ direction and one layer in the $90°$ direction. The second form is a double bias, see below.

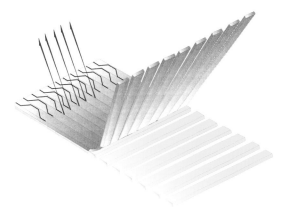

Double Bias

A biaxial reinforcement where the layers are in the $+45°$ and $-45°$ direction.

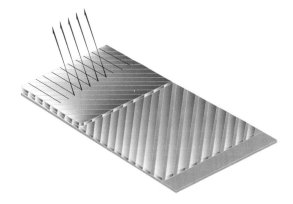

Triaxial

A three-layer reinforcement that can be either Longitudinal or Transverse:

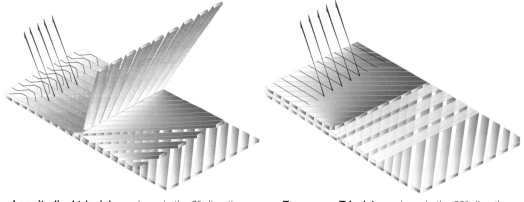

Longitudinal triaxial: one layer in the $0°$ direction and the other two layers in the $+45°$ and $-45°$.

Transverse Triaxial: one layer in the $90°$ direction and the other two layers in the $+45°$ & $-45°$.

Quadraxial

A four-layer stitched fabric with one layer in each of the four primary directions: 0°/+45°/90°/-45°.

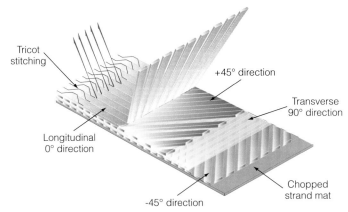

Note: All diagrams of stitched multiaxial fabrics on Pages 60–62 are courtesy of Vectorply Corporation.

Numbering System

The numbering system commonly used with these fabrics was developed by Knytex when it introduced what it called knit multiaxials in 1975. Even if a fabric is named with an unrelated product number by the manufacturer, end-users may commonly still refer to it as an 1808 or 2408, for example.

First 2 Numbers—the weight of the fabric in oz/yd^2.

Second 2 Numbers—the weight of any chopped strand mat (CSM) in oz/ft^2 x 10.

TABLE 3.12 Common Stitched Multiaxial Fabrics

Product	Fabric Weight (oz/yd²)	Mat Weight (oz/ft²)
1808	18	.8
2400	24	-
2408	24	.8
2415	24	1.5
3208	32	.8
3610	36	1.0
5600	56	-

Note: Vectorply has a 10800 product with a fabric weighing 108 oz/yd^2 and no mat.

Hybrid Fabrics

Hybrid fabrics are those made using two or more different types of fibers. Glass/aramid, carbon/aramid and carbon/glass are all common examples. There are also hybrids with carbon and Dyneema® fibers and other more exotic combinations. Hybrid fabrics may contain the different fibers as discreet layers in a stitched multiaxial or may weave two or more fiber types together in a single woven fabric.

Figure 3-6. Woven carbon/aramid hybrid twill fabric

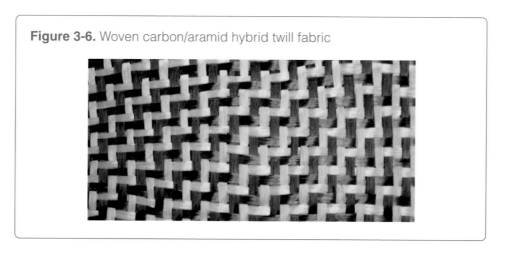

TABLE 3.13 Common Hybrid Fabrics

Manufacturer	Product	Warp Fiber	Fill Fiber	Weave
Carr Reinforcements	Style 38172	3(Carbon-3K) 1(Kevlar 158)	3(Carbon-3K) 1(Kevlar 158)	Plain
	Style 38331	1(Carbon-3K) 1(Kevlar 158)	1(Carbon-3K) 1(Kevlar 158)	3x1 Twill
	Style 38175	1(Carbon-3K) 1(Kevlar 158)	1(Carbon-3K) 1(Kevlar 158)	2x2 Twill
	Style KG390T	1(Kevlar 240) 1(Glass 600)	1(Kevlar 240) 1(Glass 600)	3x1 Twill
Fabric Development	1151	3K Carbon	1420 denier Kevlar	Plain
	4566	12K Carbon	330 yield Glass	4x4 Twill

Manufacturer	Product	Multiaxial Description	
Vectorply	KE-BX 1200	Kevlar/E-glass Double Bias	
Gurit	QEA 1201	Kevlar/E-glass Quadraxial	

Braided Fabrics

Braiding intertwines three or more yarns so that no two yarns are twisted around one another. In practical terms, braid refers to fabrics continuously woven on the bias. The braiding process incorporates axial yarns between woven bias yarns, but without crimping the axial yarns by the weaving process. Thus, braided materials combine the properties of both filament winding and weaving, and are used in a variety of applications, especially in tube construction.

Tubular braided structures feature seamless fibers from end to end and tend to perform well in torsion, which makes them well-suited for drive shafts and similar applications. In addition, braided fibers are mechanically interwoven with one another and provide exceptional impact properties as a result. Braids also provide good "channel" properties between the fiber bundles within the form that enhances resin distribution throughout a preform used in a vacuum infusion or resin transfer molding process. They are available in a wide variety of fibers, sizes and designated fiber angles.

There are 4 main types of braided fabrics:

Sleevings

These feature two sets of continuous yarns, one clockwise and the other counterclockwise, and each fiber from one set is interwoven with every fiber from the other set in a continuous spiral pattern.

Flat Braids

Flat braids use only one set of yarns, and each yarn in the set is interwoven with every other yarn in the set in a zig-zag pattern from edge to edge.

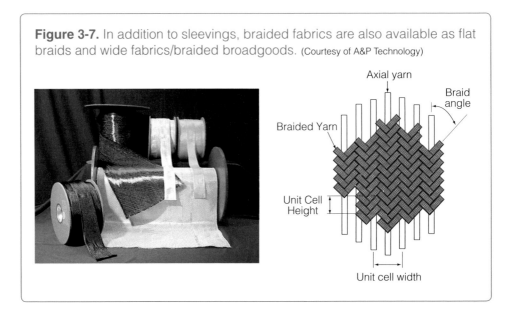

Figure 3-7. In addition to sleevings, braided fabrics are also available as flat braids and wide fabrics/braided broadgoods. (Courtesy of A&P Technology)

Wide Braided Fabrics

Also called braided broadgoods, A&P Technology produces sleevings in excess of six inches that can be slit open to create a line of wide braided fabrics. These fabrics can be produced as double bias biaxial fabrics, triaxial fabrics such as 0°, +45°, -45° or 0°, +60°, -60°, and as quadraxial fabrics, but all within one layer.

Overbraids

This is a technique where fibers are braided directly onto cores or tools that will be placed within molding processes. It is often used when a preform with very high bias angles or a contoured triaxial design is required, or when it is desirable to include circumferential windings in a preform's architecture.

There are also different types of braid architecture. Biaxial is the most common, with two yarns crossing under and over each other typically in the ±45° orientation, but also in lower angles. Triaxial braids add a third set of yarns in the axial direction, offering unidirectional and off-axis reinforcement in a single layer. (*See* Figure 3-8.)

Three-Dimensional Fabrics

These products are sometimes referred to as technical textile preforms, 3D textile preforms or complex 3D fabrics. Preforms are fiber reinforcements that have been oriented and preshaped to provide the desired contour and load path requirements of the finished composite structure. They are most commonly used in resin transfer molding (RTM), where their complex fiber orientations provide multiple flow paths for the injected resin, enabling good fiber encapsulation and higher properties. 3D fabrics can be more expensive to weave than conventional 2D woven fabrics but typically require significantly less touch labor to convert into preforms. When applied correctly, they can result in cost-effective composite structures with superior strength-to-weight ratios. (*See* Figure 3-9.)

Layup Technology

Tracers

Tracers are contrasting-colored threads woven into fabric to identify the warp and fill direction of a fabric. They are also used to show the fiber-dominant surfaces of harness satin weave fabrics. On the warp surface they are typically spaced at 2-inch intervals and run in the warp direction. On the fill surface they are typically spaced at 6-inch intervals and run in the fill direction. Tracers are not standard in all weaves and usually require a special order from the fabric vendor. In some cases tracers may be undesirable in a laminate as they are usually woven with a different type of yarn or thread. (*See* Figures 3-10 and 3-11.)

Figure 3-8. Biaxial vs. triaxial braids. (Courtesy of A&P Technology)

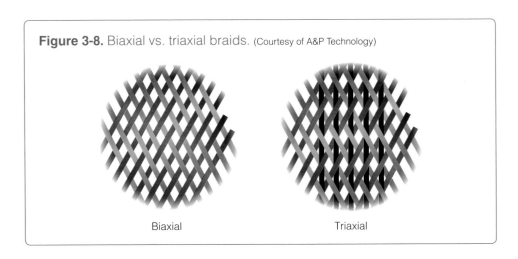

Biaxial　　　　　　　　　　Triaxial

Figure 3-9. Examples of 3D fabrics. (Photos courtesy of Albany Engineered Composites, Inc.)

(Top Left) Integrally woven turbine rotors optimize performance in hub and blade structures.

(Top Right) Flanged ring preform utilizes shape/contour woven fabric, which can be continuously wrapped with no darting or splicing.

(Bottom) 3D woven Pi preforms provide an effective means for reinforcing joints in composite structures.

Figure 3-10. Typical roll of woven fabric.

Warp fiber runs the length of the roll (blue arrow). Fill (or "weft") fiber runs across the roll and is bound at the edges with a loop stitch or a heavy selvage band (red arrow).

Figure 3-11. Tracers.

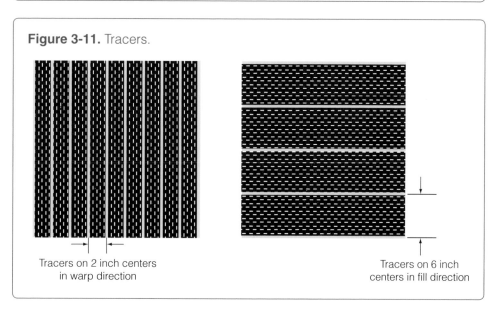

Tracers on 2 inch centers in warp direction

Tracers on 6 inch centers in fill direction

Figure 3-12. Selvage stitching at edge of roll.

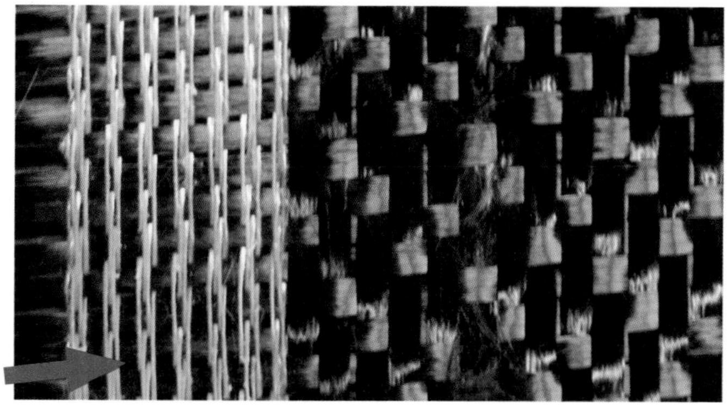

Selvage running parallel to warp fiber direction on the roll.

Selvage

The selvage is the stitched or looped edge that runs along both sides and along the length of the roll. The selvage prevents the fabric from raveling. It is typically not used in the construction of a part.

The selvage and the tracers are usually woven with a contrasting color fiber for best visibility. (Selvage can also be spelled *selvedge*, either spelling is correct.) *See* Figure 3-12 on the previous page.

Ply Kitting and Orientation Terminology

Warp Yarn Direction

Warp yarns always run the length of the roll in all conventional bidirectional woven fabrics. The warp yarn direction of the fabric is always used for orientation purposes in composite panel manufacturing.

Fill Yarn Direction

Fill yarns run across the roll, $90°$ to the warp yarns. These yarns are also called "weft" yarns in the textile industry.

Warp Face

Harness satin weave styles have one surface or "face" in which the most-visible yarns are running in the warp direction. The opposite face has the most-visible yarns running in the fill direction. The side that is warp-dominant is called the "warp face" of the fabric. The warp face of a harness satin fabric is always identified, and is positioned either "warp up" or "warp down" in a layup, according to the laminate design. Warp face recognition is critical to proper **nesting** and **symmetry** in a layup using harness-satin weave fabrics, and to panel stiffness properties in the laminate design.

Yarn Count

The "yarn count" is typically listed in yarns per inch. Woven fabrics that have an equal number of identical yarns in both the warp and fill directions are called symmetric-weave fabrics and will be, for all practical purposes, equally strong in both directions.

Symmetric plain or twill weave fabrics can often be oriented at either $0°$ or $90°$ and have the same properties in each direction. However, symmetric **harness-satin weave** fabrics inherently have a warp face and a fill face that are essentially $0°$ dominant on one side and $90°$ dominant on the other. Although they can be oriented at either $0°$ or $90°$ in a layup (or at other angles) and will have the same strength in each direction, the dominant face orientation of the outer surfaces of the laminate will have a greater flexural stiffness on the axis of the outer-most fibers. Thus, unlike symmetric plain weave fabrics, which have equal flexural stiffness in both directions, symmetric harness-satin weave fabrics have dominant fiber directions at each face and cannot be oriented at either 0° or 90° with equal flexural stiffness properties in each direction.

A woven fabric that has a different number of yarns per inch in the warp and fill directions, or different size yarns in the two directions, or different types of yarns in the two directions, or combinations of the above, may be called asymmetric-weave fabrics. It will not necessarily have equal strength in both directions. Typically, asymmetric-weave fabrics have a greater number of warp yarns than fill yarns per inch and thus necessitate specific orientation requirements.

Basic Design Considerations

4

In this chapter:
- Composite Structural Design
- Fiber Orientation
- Resin to Fiber Ratio
- Service Life Considerations

Composite Structural Design

Matrix-Dominated Properties

The following are all matrix-dominated properties:

- Compressive strength
- Off-axis bending stiffness
- Interlaminar shear strength
- Service temperature

Damage to the matrix system, delaminations, porosity, or too low of a resin/fiber ratio will cause these properties to degrade rapidly. Matrix damage or degradation has a small effect on the tensile strength of a composite structure, yet at the same time it has a significant effect on compressive strength and shear properties.

Fiber-Dominated Properties

- Tensile strength
- Flexural modulus

Tensile strength and flexural modulus (stiffness) are fiber-dominated properties. The type of fiber reinforcement used, the fiber form that is selected, and the orientation arrangement of the fibers have a strong effect on the laminate performance.

Solid Laminate Panels

Solid laminate panels are typically built using either woven cloth, unidirectional tape, or a combination of both. Unidirectional tape yields the highest strength-to-weight ratios, and the most efficient on-axis load transfer, but typically requires

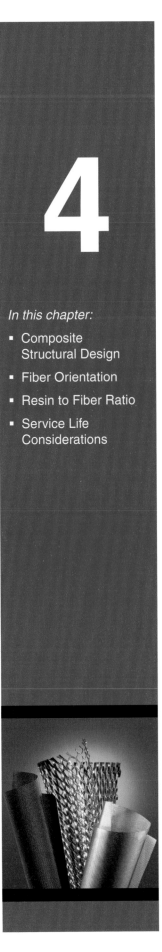

71

more plies to achieve quasi-isotropic, balanced, and symmetric properties (discussed later in this chapter). Fiber placed unidirectional forms function much like unidirectional tape.

These types of laminates are best used where high flexural, tensile, and/or compressive loads are required. They are often seen in wing skin structures, as well as spars, stringers, floor beams, and other stiffeners. Stiffener designs include *I-beam*, *blade* and *hat-section* stiffeners (*see* Chapter 6, Page 118), named for their shapes, and preferably created as an integral part of the structure. Secondarily-bonded stiffener structures tend to be less efficient and more labor intensive to build.

The specific performance properties of these structures are dependent on the fiber or ply orientations, especially with unidirectional forms. Woven cloth ply orientation is also important, but is usually uniformly bi-directional. Non-woven discontinuous fiber "mat" materials are not at all dependent on orientation, and are quasi-isotropic by their nature; however they cannot match the strength or stiffness of oriented continuous or long-fiber reinforced laminates.

Solid-laminate structures tend to be more durable and damage tolerant than sandwich panels, which are discussed below.

Sandwich Core Panels

"Sandwich" structures rely on relatively thin skins bonded to a lightweight but usually thick core material. Compared to solid laminates, they offer much higher bending stiffness-to-weight ratios. They are primarily used in applications in which high stiffness-to-weight is required (example: helicopter rotor blades, landing gear doors, control surfaces, and floor panels).

The primary disadvantage of sandwich structures is their reduced damage tolerance and durability compared to solid laminates. The thin skins can often be scratched or punctured. In addition, water or other liquids can ingress into honeycomb cores, becoming trapped. These liquids can be difficult to locate and remove and can also cause disbonding of the skins when the liquid freezes and expands, or when subjected to temperatures above the boiling point during service or repairs.

Temperature and Moisture

The wet T_g of the matrix resin dictates the maximum service temperature that the structure can take without degrading. Typically the higher the moisture content absorbed into the laminate, the lower the service temperature will be. This is especially notable in laminates and sandwich structures that are exposed to high temperatures surrounding engine and exhaust areas. Good **"hot-wet"** properties of the combined matrix/fiber structure can be examined via thermal analysis and destructive coupon testing.

Ply Orientation and Standardized Orientation Symbol

Since the laminate performance is almost totally dependent upon fiber orientation, this must be closely controlled at the engineering, manufacturing, and repair levels. Because of this, there is an internationally recognized system for controlling fiber orientation on engineering drawings and all subsequent manufacturing processes. This involves the use of an orientation symbol and a specified location on the structure where the fiber orientation is prescribed and controlled. The fiber angle callouts are usually tabulated and coordinated to the ply orientation symbol/location.

The recognized standard for use of this symbol states that the counter-clockwise (CCW) symbol as shown in Figure 4.1 will be used on the main views on the composite panel manufacturing drawing to show the relative fiber or warp yarn direction, normally with the 0° axis parallel to the primary load direction of the laminate structure or assembly.

In addition, the CCW symbol is always shown on the inside or bag-side surface of the part, as looking towards the primary tool surface, exactly where it is applicable. The fiber orientation will be controlled and inspected at this exact location in manufacturing. This is especially critical on contoured parts, whereas the fiber direction may change drastically as the fibers/fabrics are draped over the contoured shape. The location of the symbol is less critical on flat panels.

Multiple symbols may be used on large or complex parts if necessary. Coordination of the applicable symbol/location with the ply layup table (or tables) is necessary when multiple symbols are used.

Normally a fiber alignment tolerance is also specified in the general drawing notes that coincide with the location/position of this symbol (e.g., ± 2° to ± 5°).

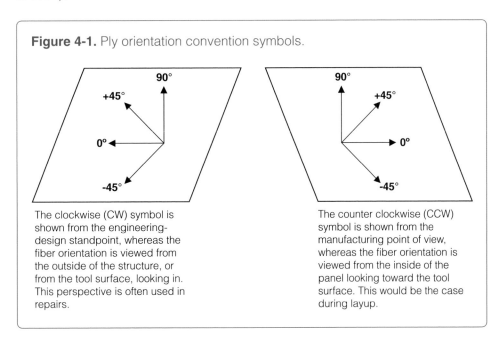

Figure 4-1. Ply orientation convention symbols.

The clockwise (CW) symbol is shown from the engineering-design standpoint, whereas the fiber orientation is viewed from the outside of the structure, or from the tool surface, looking in. This perspective is often used in repairs.

The counter clockwise (CCW) symbol is shown from the manufacturing point of view, whereas the fiber orientation is viewed from the inside of the panel looking toward the tool surface. This would be the case during layup.

Ply Layup Table

The ply layup table is used on the composite panel drawing to define the ply layup sequence, ply numbers or part numbers, orientation requirements, materials, **splice** control, and other pertinent information about the panel layup. An example is shown in Table 4.1.

TABLE 4.1 Ply Layup Table

Sequence	Ply Number	Orientation	Material	Splice Control	Revision
010	P1	+45	A⟩	6⟩	-
020	P2	-45	A⟩	6⟩	-
030	P3	0	A⟩	6⟩	-
040	P4	90	A⟩	6⟩	-
050	P5-P8	0	C⟩	9⟩	-
060	P9	90	A⟩	6⟩	-
070	P10	0	A⟩	6⟩	-
080	P11	-45	A⟩	6⟩	-
090	P12	+45	A⟩	6⟩	-

Common Layup Terms and Conditions

Symmetry

A *symmetric* laminate is one in which the ply orientations are symmetric about the middle (or *mid-plane*) of the laminate when viewed in cross-section. Another way of looking at this is to describe the ply orientations of the laminate as being a mirror image from the mid-plane. Laminate symmetry is necessary to maintaining dimensional stability through processing and in service. (*See* Figure 4-2, 4-3.)

Achieving a symmetric laminate is often difficult with harness satin fabrics and stitched-biaxial or multiaxial forms. A thorough understanding of how the fiber/fabric forms can affect laminate symmetry is necessary when creating properly designed laminate structures that will maintain shape when processed and/or when exposed to changes in temperature in service.

Balance

A *balanced* laminate refers to a laminate with equal numbers of plus- and minus-angled plies. This arrangement helps to avoid twisting under applied loads. Typically a laminate that is both balanced and symmetric will have an equal number of plus- and minus-angled plies on each side of the laminate mid-plane. (*See* Figure 4-4 on the next page.)

Figure 4-2. Symmetry in layups.

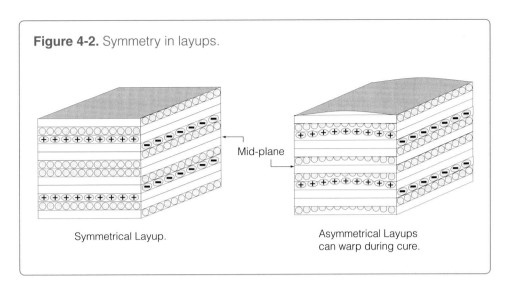

Mid-plane

Symmetrical Layup.

Asymmetrical Layups can warp during cure.

Figure 4-3. Effect of symmetry in 0° and 90° laminates.

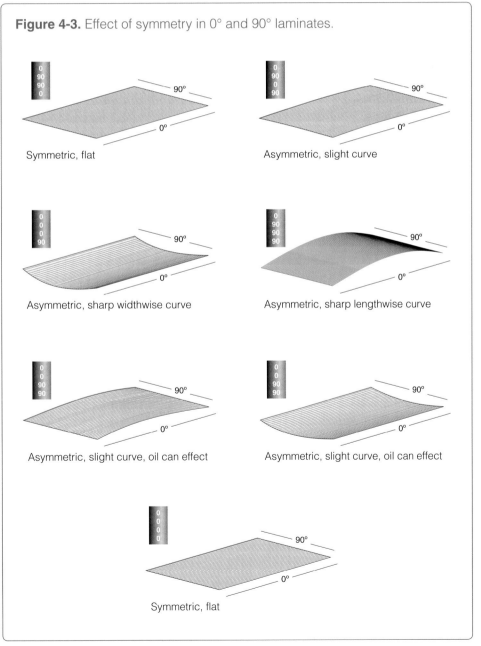

Symmetric, flat

Asymmetric, slight curve

Asymmetric, sharp widthwise curve

Asymmetric, sharp lengthwise curve

Asymmetric, slight curve, oil can effect

Asymmetric, slight curve, oil can effect

Symmetric, flat

Figure 4-4. Effect of balance and symmetry in +45° and -45° laminates.

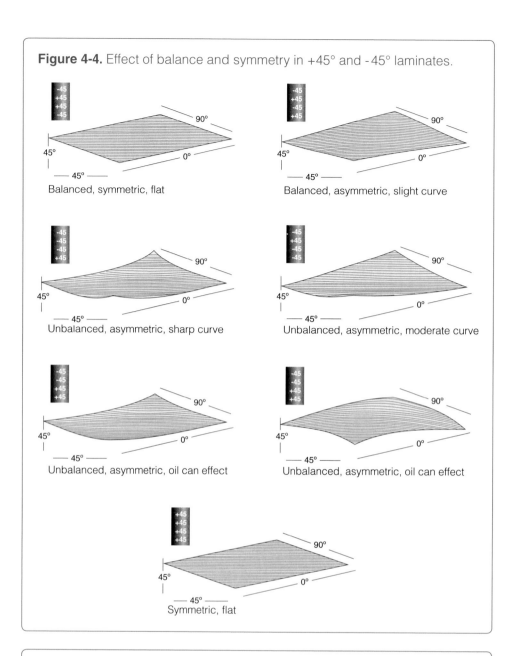

Balanced, symmetric, flat

Balanced, asymmetric, slight curve

Unbalanced, asymmetric, sharp curve

Unbalanced, asymmetric, moderate curve

Unbalanced, asymmetric, oil can effect

Unbalanced, asymmetric, oil can effect

Symmetric, flat

Figure 4-5. Quasi-isotropic laminates.

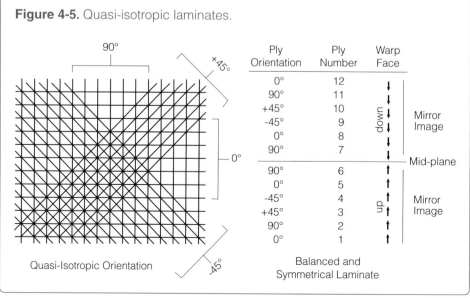

Quasi-Isotropic Orientation

Ply Orientation	Ply Number	Warp Face	
0°	12	↓	
90°	11	↓	
+45°	10	↓ down	Mirror Image
-45°	9	↓	
0°	8	↓	
90°	7	↓	
			Mid-plane
90°	6	↑	
0°	5	↑	
-45°	4	↑ up	Mirror Image
+45°	3	↑	
90°	2	↑	
0°	1	↑	

Balanced and Symmetrical Laminate

Quasi-isotropic

This term simply means in-plane isotropic properties. A quasi-isotropic composite part has either discontinuous (chopped) fiber in a random orientation, or has continuous (long) fibers oriented in directions such that equal strength is developed all around the laminate when loaded in edgewise tension or compression.

Generally, a laminate is considered to be quasi-isotropic when laid up with an equal number of plies at each of $0°$, $90°$, $+45°$, and $-45°$ angles. Quasi-isotropic properties can also be achieved with fewer fiber angles (e.g. $0°$, $+60°$ and $-60°$), or with additional fiber angles (e.g. $0°$, $+22.5°$, $+45°$, $+67.5°$, $90°$, $-67.5°$, $-45°$ and $-22.5°$). Obviously the first example cannot be achieved with bidirectional fabrics, and in the second example more plies would be required to accommodate this configuration. Thus, the industry standard is to use the $0°$, $90°$, $+45°$ and $-45°$ angles. This is compatible with both unidirectional and bidirectional material forms.

Nesting vs. Stacking

Nesting refers to the placement of harness-satin woven cloth plies so that the dominant fiber direction at the surface of one ply aligns in the same direction as the interfacial surface of the adjacent ply. The yarn direction at this interface is the same for each ply at this interface, thus the plies are "nested" together.

Stacking refers to the placement of plies in the layup without regard for up/down positioning of the warp face. Plain weave and twill weave fabrics are typically stacked as they do not have a visible warp or fill-dominant face on which to refer.

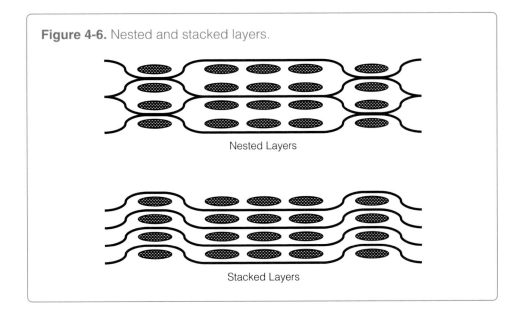

Figure 4-6. Nested and stacked layers.

Nested Layers

Stacked Layers

- If the two plies in question will have the same warp fiber orientation, orient them warp face to warp face, or fill face to fill face to achieve a nested interface.

- If the two plies in question have warp fiber orientations $90°$ apart, orient them warp face to fill face and they will nest.

- If the two plies in question are oriented at angles other than the same or $90°$ to each other, nesting is not possible; therefore, a +45 or -45 angled ply cannot nest with a $0°$ or $90°$ ply, etc.

Considerations with Symmetric Fabrics

With a symmetric plain weave cloth, the two faces are identical to each other and have the same number of identical yarns running in both the warp and fill direction. Therefore $0°$ or $90°$ plies are essentially the same thing, and are called 0/90s. The same goes for +45 and -45 plies, which are also called ±45s.

- Balance naturally applies, since every $±45°$ ply automatically has both ±45 fiber/yarn directions, and so is inherently balanced.

- Symmetry in terms of ply orientation sequence from the mid-plane is still required to make a dimensionally stable laminate, although it can be achieved with fewer plies. Note that there must always be an even number of harness-satin plies to achieve symmetry in any case because of the warp/fill face dominance.

- A layup would still have equal numbers of 0/90 and $±45°$ woven fabric plies to qualify as quasi-isotropic.

- Nesting does not apply to a plain weave fabric laminate.

- If working with a symmetric harness satin fabric, attention to the warp and fill face orientation would still be required in order to achieve symmetry.

Considerations with Asymmetric Fabrics

A good example of an asymmetric harness satin fabric is style #1581 (or #7781) glass, in which the yarn count for these fabrics is 57 warp yarns to 54 fill yarns per inch within the weave pattern. Thus, the warp direction of the fabric has a greater fiber count and is slightly stronger than the fill direction of the fabric. The fabric cannot be laid-up 0/90 or ±45 without recognition of the effect on symmetry and structural loading.

All of the above conditions apply to unidirectional tape laminates, except that nesting applies only when multiple layers are stacked running in the same orientation. Since there are no warp or fill faces, the typical rules of nesting as listed above are not applicable. Since uni-tape is unidirectional, it is inherently an asymmetric layer.

Ply Orientation Shorthand Code

Shorthand codes are used during engineering analysis or for other non-drawing uses. They are a way to condense a long descriptive representation of a multi-layer laminate into as few symbols as possible.

General Laminate Description

- Each lamina is identified by its ply fiber angle callout. No degree symbol is used within the shorthand brackets.

- Lamina are listed in sequence starting from the tool surface or surface indicated by the leader arrow.

- Each lamina is separated by a slash.

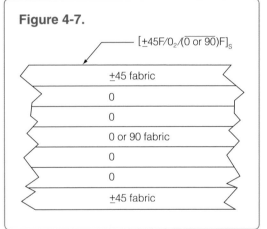

Figure 4-7.

$[\pm 45F/0_2/(\overline{0 \text{ or } 90})F]_S$

+45 fabric
0
0
0 or 90 fabric
0
0
+45 fabric

- Multiple laminae of the same angle are indicated by a subscript, indicating the number of plies, following the angle indication.

- Repeating groups of plies or special entities within a laminate can be placed in parenthesis.

- Each laminate shorthand code is enclosed in a set a brackets.

- Symmetric laminates with an even number of plies are represented by listing all plies on one side of the mid-plane enclosed in brackets, followed by a subscript S.

- Symmetric laminates with an odd number of plies are listed with a line over the center ply to indicate that it is the mid-plane enclosed in brackets, followed by a subscript S.

Unidirectional Laminae

- ±45 indicates two unidirectional plies starting with a +45 followed by a -45.
- 45 indicates two unidirectional plies starting with a -45 followed by a +45.

Fabric Laminae

- Fabric plies are identified with a capital F following the ply angle callout.
- The angle value represents the direction of the warp fibers.
- ±45F indicates a woven fabric placed at either a + or - 45 degree direction.
- +45F indicates that the warp fiber must be placed in the +45 direction.
- (0 or 90)F indicates a fabric ply at either 0 or 90 degrees.

Hybrid Laminates

- Individual plies of different materials located in a laminate are identified by a subscript following the ply angle identifying the type of material.
- When numerous plies of the same material type are grouped together, the material subscript is shown on only the first and last ply in the group. Everything in between is assumed to be the same material.

Figure 4-8.

$[\pm 45F/0/90]_S$

+45 Uni-tape
-45 Uni-tape
0 Uni-tape
90 Uni-tape
90 Uni-tape
0 Uni-tape
-45 Uni-tape
+45 Uni-tape

Figure 4-9.

$[\pm 45F_{FG}/(0/90)_{2C}/0_C]_S$

±45 fiberglass fabric
0 uni-carbon
90 uni-carbon
0 uni-carbon
90 uni-carbon
0 uni-carbon
0 uni-carbon
90 uni-carbon
0 uni-carbon
90 uni-carbon
0 uni-carbon
±45 fiberglass fabric

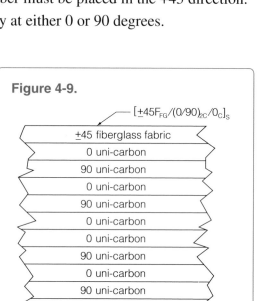

Resin to Fiber Ratio

For advanced composite laminates, there is a fairly narrow range of functional resin/fiber ratios which will produce the best strength-to-weight ratio in the end-product. With too little resin, compressive and shear properties drop-off. With too much resin (assuming all of the fiber is still in the composite), the laminate's weight and stiffness go up without any significant increase in strength. Thus, the strength-to-weight ratio goes down.

With woven fabric forms in the laminate, the target fiber to resin percentage is approximately 60% fibers to 40% resin, by volume. In the shop environment, a "by weight" percentage is usually easier to calculate. However, it should be noted that a 60/40 weight percentage does not necessarily equal a 60/40 volume fraction for all fiber/resin combinations. There will be many variations of this fiber/resin weight relationship, depending on the type of fiber, form, and the type and amount of resin used. See the example in Table 4.2 below.

Unidirectional tape forms typically require a lower resin content, as there is not much area between the fibers that need to be filled with resin like there is with a woven form. Resin ratios in the low to mid 30% range, by volume, yield the best results (e.g., 31% resin/ 69% fiber).

TABLE 4.2 Calculating Resin to Fiber Ratio

By Weight	By Volume
Wet layup example. Ideal ratio in this example is 40% resin/60% fibers, by weight.	This side illustrates just how different the actual fiber-volume could be if the weight ratio is used to make the laminate.

By Weight	By Volume

Calculating Resin Weight
 	
cloth weight ÷ .6 (60%) = total weigh
 	
total weight - cloth weight = resin weight
 	
Example: 90 g cloth ÷ .6
 	
    = 150 g total weight
 	
        150 - 90 = 60 g resin weight
 	
Calculating Actual Ratio
 	
actual total weight - cloth weight
 	
= resin weight
 	
resin weight ÷ total weight = resin %
 	
Example:
 	
150 g actual weight - 90 g cloth
 	
= 60 g resin
 	
60 ÷ 150 = .40 (or 40%) resin ratio
 	
Therefore, panel's resin-to-fiber by weight ratio is 40/60, if only 60 grams of resin is present in the laminate.
	Fiber Weight = 90 g
	
	Composite Weight = 150 g
	
	Resin Weight = 60 g
	
	Fiber Density = 2.65 g/cc
	
	Resin Density = 1.24 g/cc
	
	Fiber Volume = 33.96 cc
	
	Fiber/Resin Weight Ratio = 1.5
	
	Fiber/Resin Volume Ratio = 0.7
	
	Resin Volume = 58.8%
	
	Void Volume = 1.0%
	
	Fiber Volume = 40.2%
	
	Therefore, excluding void volume, the panel's resin to fiber volume ratio is about 60/40, just the opposite of the by weight ratio shown in the previous example.

Environmental Effects on Cured Composite Structures

Fuels and Oils

Typical fuels and oils, either gasoline or jet fuels, have little if any effect on cured epoxy resin, in terms of chemical attack. However, they can easily penetrate a porous skin and saturate underlying core materials. While this effect may not directly weaken the part, it in many cases eliminates any possibility of a repair cured under vacuum, as the fuel or oil would be drawn into the bondline with vacuum and contaminate the repair.

Hydraulic Fluids

Military hydraulic fluids, specifically 5606 and 83282 apparently do not have a direct chemical effect on most cured resins. Of course, the contamination effect described above can easily take place with any fluid ingress. Some types of Skydrol® brand hydraulic fluids can attack cured epoxy, and weaken it. These fluids are used primarily in civilian aircraft and airline equipment.

Deicing/Anti-Icing Fluids

The risk of fluid ingress into the composite structure is the biggest risk associated with deicing fluids. Otherwise, deleterious effects due to chemical attack have not been documented to date.

Acids/Alkalis

Weaker acids, such as battery acid, do not have any harmful effects on cured epoxy. However, battery acid can attack glass fibers and chemically etch them. This is especially true with E-glass. Therefore, C-glass is used to minimize acid attack problems. Strong acids, such as might be found in a chemical laboratory, can attack cured resins and should be isolated from composite structures.

With the exception of nickel-cadmium (Ni/Cd) electrolytes, weaker alkalis normally do not attack cured resins. However, they can easily attack some of the structural foams used in sandwich panels, although this effect varies substantially depending on the type of foam core material. In some cases, even ordinary soaps or detergents can dissolve some types of foam cores. Some types of glass fibers are also subject to alkali attack.

Solvents

Most of the ordinary solvents found in workshops do not harm cured epoxies. However, one particular type of solvent, Methylene Chloride, will chemically attack cured epoxy resin, and cannot be used on composite structures.

(Please note that this is an entirely different issue than any health and safety considerations these materials may have.)

Paint Strippers

Chemical paint strippers are largely based on Methylene Chloride, and therefore cannot be used on composites. Other paint stripping methods must be used, which are discussed in Chapter 10.

Ultraviolet

Ultraviolet (UV) light, primarily from sunlight, will slowly attack cured resins and cause the matrix to become brittle and prone to cracking. In addition, aramid fibers are also subject to weakening with prolonged UV exposure. One can easily protect composites from UV by using an opaque paint with a UV barrier.

Infrared

Infrared light is heat radiation, and is a particular problem with the bottom side of wings and tail surfaces while the aircraft are parked on a black asphalt ramp on a hot summer day. These surfaces receive direct infrared radiation from the ramp. Temperatures of 225°F/107°C have been measured in such conditions, while air temperatures were approximately 120°F/49°C. Protect the structures by painting them white on all lower surfaces and this will reflect away most of the infrared radiation.

Galvanic Corrosion

Galvanic corrosion occurs when two materials from different positions on the galvanic scale come in direct contact with each other, and is aggravated by the presence of moisture. Dissimilar metal corrosion problems are well known. However, carbon fiber composites also exhibit this galvanic effect with some metals. The corrosion occurs in the metal alone, as the carbon fiber composite itself is unaffected. (*See* Table 4.3 on the next page.)

To avoid corrosion, the carbon fiber composite must be isolated from the metal with a suitable dielectric material. This is often accomplished with a thin layer of fiberglass as an interface ply, or a layer of a non-conductive film adhesive/carrier at the interface.

Mechanical fasteners must also be selected with this problem in mind. Severe fastener corrosion is not uncommon in carbon fiber composites. Cad-plated fasteners, for example, can corrode rather quickly in such structures.

TABLE 4.3 Corrosion Potential of Various Metals with Carbon Fibers

Material	Galvanic Scale	
Magnesium	12	
Zinc	11	
Aluminum 7075 Clad	10	ANODIC
Aluminum 2024 Clad	9	
Aluminum 7075-T6	9	
Cadmium	8	
Aluminum 2024-T4	7	
Wrought Steel	6	
Cast Steel	6	
Lead	4	
Tin	4	
Manganese Bronze	3	
Brass	2	
Aluminum Bronze	2	
Copper	2	
Nickel	1	
Inconel	1	
300 Series 18-8 CRES	0	
Titanium	0	
Monel	0	CATHODIC
Silver	0	
Carbon Fiber	0	

Lightning Strike Protection (LSP)

Note: Text from this section is adapted from "Lightning Strike Protection For Composite Structures" by Ginger Gardiner, *High Performance Composites Magazine*, July 2006; reproduced by permission of *High Performance Composites Magazine*, copyright 2006, Gardner Publications, Inc., Cincinnati, Ohio, USA.

Unlike metal structures, parts made from composites do not readily conduct away the extreme electrical currents and electromagnetic forces generated by a lightning strike. Composite materials are either not conductive at all (e.g. fiberglass) or are significantly less conductive than metals (e.g. carbon fiber). When lightning strikes an unprotected composite structure, up to 200,000 amps of electrical current seeks the path of least resistance to ground, usually finding any metal available. Along the way, it can cause a significant amount of damage. It may vaporize metal control cables, weld hinges on control surfaces and explode fuel vapors within fuel tanks if current arcs through gaps around fasteners. These direct effects also typically include vaporization of resin in the immediate strike area and possible burn-through of the laminate. Indirect effects occur when magnetic fields and electrical potential differences in the structure induce transient voltages, which can damage and even destroy any electronics that have not been EMF (electromagnetic field) shielded or lightning protected. For these reasons, lightning strike protection (LSP) is a significant concern, especially in vulnerable composite structures such as sailboat masts, aircraft and wind turbines.

LSP Design Fundamentals

LSP strategies have three goals:

1. Provide adequate conductive paths so that lightning current remains on the structure's exterior.

2. Eliminate gaps in the conductive path to prevent arcing at attachment points and possible ignition of fuel vapors.

3. Protect wiring, cables and sensitive equipment from damaging surges or transients through careful grounding, EMF shielding and application of surge suppression devices where necessary.

Traditionally, conductive paths in composite structures have been established in one of the following ways: (1) bonding aluminum foil to the structure as the outermost ply; (2) bonding aluminum or copper mesh (expanded foil) to the structure either as the outermost ply or embedded within a syntactic surfacing film; or (3) incorporating strands of conductive material into the laminate.

All of these approaches require connecting the conductive pathways to the rest of the structure in order to give the current an ample number of routes to safely exit to ground. This is typically achieved by using metal bonding strips (i.e., electrical bonding) to connect the conductive surface layer to an internal "ground plane," which is usually a metal component such as an aircraft engine or wind turbine motor, a metal conduit pipe, a metal base plate in a sailboat, etc. Because lightning strikes can conduct through metal fasteners in composite structures, it may be desirable to prevent arcing or sparking between them by sleeving countersunk holes with conductive sleeves that contact the primary LSP, or encapsulating fastener nuts or sleeves with plastic caps or polysulfide coatings.

Figure 4-10. Microgrid® expanded metal LSP.

Dexmet supplies a large variety of conductive metal products for aircraft, including aluminum, copper, phosphorous bronze, titanium and other materials. (Photo courtesy of Dexmet Corporation)

LSP Materials

The need for protection of composite structures has prompted development of a number of specialized LSP materials.

Metal Mesh/Foil Products

For external surface protection, thin metal alloys and metallized fibers have been developed into woven and non-woven screens and expanded foils. These mesh-like products enable lightning current to quickly travel across the structure's surface, reducing its focus. Dexmet supplies a large variety of conductive metal products for aircraft, including aluminum, copper, phosphorous bronze, titanium and other materials. Dexmet can modify any design to meet precise customer needs and works with customers to test and evaluate LSP designs. Dexmet can also supply some LSP in 48" widths, which can reduce labor cost during application.

Aluminum wire was one of the first LSP materials used, interwoven with carbon fiber as part of the laminate. However, using aluminum with carbon fiber risked galvanic corrosion, so copper wires were then tried, but are three times as heavy as aluminum. As the aircraft industry began using more fiberglass composites in aircraft, it investigated foils and then expanded foils, which can be co-cured with the laminate's exterior ply.

Coated fibers (nickel or copper electrodeposited onto carbon and other fibers) are also used but perform much better as EMF shielding than as direct lightning strike protection. Hollingsworth & Vose has developed a non-woven veil from nickel-coated carbon fiber. It is flexible, lightweight and has performed well in lightning strike testing, with no damage to any fibers below the LSP.

LSP Prepregs

"All-in-one" LSP prepregs contain pre-embedded woven or non-woven metal meshes and are applied first-down in layups. By combining the LSP and external prepreg surface layers into one ply, these products may reduce kitting and manufacturing costs.

Cytec produces SURFACE MASTER 905® with expanded copper screen (ECS) or expanded aluminum screen (EAS) already embedded in, specifically designed to produce a composite structure with no pinholes. Somewhat pliable and drapeable like a film adhesive, SURFACE MASTER 905® contains enough resin to permit surface sanding without damaging the embedded metal screen. It is available in several areal densities and foil thicknesses for weight optimization.

Loctite Hysol also produces a prepreg surface film that goes by the brand name SYNSKIN®. This material is available with the expanded foil in different densities to meet specific structural applications.

Post-manufacture External Products

Integument Technologies produces polymer-based, peel-and-stick appliqués, which can be installed on composite surfaces after construction. Lightning Diversion Systems makes a thin *conformal shield* product for composite exterior surfaces, which protects against both direct and indirect lightning damage, is lightweight and produces a smooth finish. Vought Aircraft and Kaman Aerospace have used the product, which is more expensive than machined metal scrims but also can be stretched, providing enhanced conformability to structures with complex curvature.

LSP Issues

Paint Thickness

Adam Aircraft used Astrostrike non-woven copper mesh in the carbon fiber composite airframes of their A500 and A700 aircraft. They primed and painted as usual, but later learned that the paint thickness must be kept to a specified minimum. If the paint is too thick, lightning does not readily conduct through to the copper mesh, which basically prevents the LSP from working.

Porosity

After a lightning strike, repairs must be done properly to regain the conductive path. Traditional methods of repairing LSP-enabled composite structures using metal mesh and film adhesive often result in surface porosity that later permits moisture ingress. However, using a mesh-embedded surfacing film can reduce or actually eliminate surface porosity. The optimum surface is obtained by using separate layers of mesh and surfacing film, applied so that the film overlaps the mesh and provides an adequate fillet around the repair edges. Compared to previous methods, this approach replaces the outermost layer of film adhesive with surfacing film in order to bond and seal the mesh-to-mesh interface, so that the repair patch mesh is bonded and sealed to the original surface mesh, resulting in a successfully restored conductive path.

Next Generation LSP

Current research with conductive nano-particle coatings, primers, paints, and fillers will greatly enhance future LSP capabilities. These materials are currently under development and will eventually complement today's materials with much more efficient lightning strike protection capabilities.

5

Molding Methods and Practices

Overview of Molding Methods and Practices

There are many different molding methods and techniques employed in the manufacture of advanced composite parts. Some methods use a single mold surface with a vacuum bag on the backside of the layup, and some use a cavity mold or die-set. The material selection, complexity of the shape, and production rate requirements often drive the choice of molding methods for a specific composite structure.

There are two basic categories of composites processing: open molding and closed molding. In open molding, the laminate is exposed to the atmosphere during the fabrication process. In closed molding, it is either processed using a two-part mold which shapes the external and internal surface, or within a vacuum bag. A wide variety of methods exists within these two categories as can be seen in Table 5.1 below.

TABLE 5.1 Open and Closed Molding Processes

Open Molding	Closed Molding
• Hand Layup	• Vacuum Bagging
• Spray-up	• Prepreg Layup (Automated Tape Laying)
• Filament Winding	• Vacuum Infusion
	• Resin Transfer Molding (RTM)
	• Pultrusion
	• Matched Die Forming
	• Compression Molding

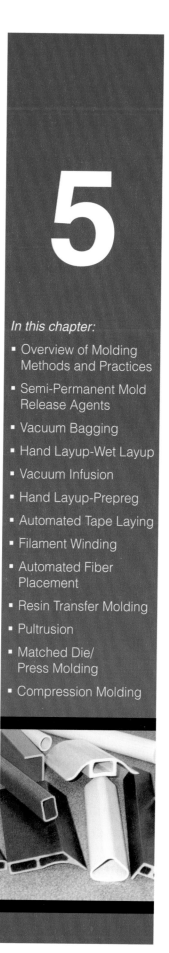

In this chapter:
- Overview of Molding Methods and Practices
- Semi-Permanent Mold Release Agents
- Vacuum Bagging
- Hand Layup-Wet Layup
- Vacuum Infusion
- Hand Layup-Prepreg
- Automated Tape Laying
- Filament Winding
- Automated Fiber Placement
- Resin Transfer Molding
- Pultrusion
- Matched Die/ Press Molding
- Compression Molding

Molding liquid resin and dry fibers often demands a different approach than molding pre-impregnated (prepreg) materials, although some parts of these processes are similar. The primary difference involves the mixing and distribution of A-staged resin throughout the fibers versus having the resin already pre-impregnated in the fiber/fabric layers. With wet-layup "contact molding" processes, a surface coat of resin is usually applied to the mold prior to the wet layup and the plies are either "wet-pregged" on a covered table and transferred to the mold or interleafed with alternating resin/fiber/resin/fiber sequences. This technique generally results in uneven and/or inconsistent resin content in the end product. Strict controls on resin amounts and distribution can alleviate some of the inconsistencies.

With resin infusion processes, a dry-pack or preform is placed in the mold and the resin is introduced through an injection process using pressure and/or vacuum. Sometimes a surface coat of resin is pre-applied to the mold prior to the installation of the preform. These processes use either a single-sided mold with a vacuum bag, as with a Vacuum Infusion Process (VIP), or a multi-piece cavity mold, as with Resin Transfer Molding (RTM) processes.

Prepregs are used in a variety of different and sometimes unique molding processes ranging from contact molding in a single-sided mold with a vacuum bag and oven or autoclave cure, to closed molding processes.

With the advent of Computer Aided Design (CAD) and Computer Aided Manufacturing (CAM), many methods have evolved from previously labor intensive hand-placed material processes to more streamlined automated processes. Pultrusion, filament winding, automated tape laying, and fiber placement technologies are a few of the more common processes that have evolved since the 1960s. For example, today's 6-axis fiber-placement process was spawned from yesterday's 2-axis filament winding technology. Many new hybrid processes will develop as the industry finds more innovative ways to efficiently produce fiber-reinforced plastic parts.

Semi-Permanent Mold Release Agents

In molding and assembly of composite parts, high performance release agents are used for releasing parts from sticking to the mold or fixture. Much like a Teflon® coated pan, semi-permanent mold release systems are in fact cured polymers that are bonded to the tool surface. (They are not wax!) Typically, a semi-permanent system is made up of two different polymers:

1. Tool sealer coat to fill the micro-porous sites on the tool surface (reducing mechanical attachment).

2. Low-energy release agent to prevent chemical attachment (*see* Figure 5-1).

These materials work best when applied in very thin layers. Application procedures generally require that the materials be sprayed or wiped sparingly onto the tool surface. The desired result is to have the least amount of release agent on the surface that will do the job. This reduces the possibility that the release polymer will transfer to the part surface during the molding opera-

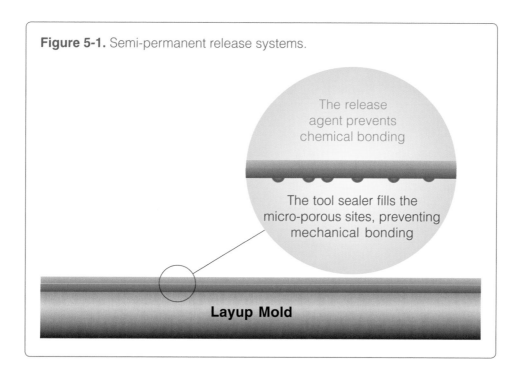

Figure 5-1. Semi-permanent release systems.

The release agent prevents chemical bonding

The tool sealer fills the micro-porous sites, preventing mechanical bonding

Layup Mold

tion. Worse yet, the release polymer can act like an adhesive when it is over-applied and/or under-cured. This sometimes results in the part sticking to the tool and perhaps damaging both the part and the tool upon demolding. Thus, proper application procedures are crucial for good performance.

A semi-permanent release agent is designed to provide multiple releases. Typically 2–10 parts can be molded from a properly treated surface prior to re-application of the release agent. Some manufacturers choose to re-apply the release coat every molding cycle. While this is thought to ensure a good release, it actually adds to quicker build-up and possible transfer problems in production.

After a certain number of applications it is recommended that the release agent be stripped back to the sealer and reapplied. This is necessary for good long-term release performance without release transfer. Most manufacturers will set up regular maintenance intervals of 20–50 application cycles, depending on the tooling and the release system used.

Semi-permanent release systems are available in both solvent and solvent-free formulas. Many manufacturers are choosing to use the solvent-free, non-**VOC** formulas in lieu of the old solvent-based materials in order to eliminate the health and safety risks and reduce the environmental concerns that they cause.

Vacuum Bagging

The primary function of a vacuum bag is to provide compaction pressure and consolidation of the plies within the laminate. This is done by creating a pressure differential between the inside and the outside of the bag using a vacuum pump, a hose, and a sealed film membrane (bag), and results in down-force on the bag equal to the ambient atmospheric pressure.

The secondary function of a vacuum bag, used along with a bleeder/breather mechanism for processing outside of the autoclave, is to facilitate removal of excess resin and/or gases from a laminate during processing. The routes for extraction of gases and resins are via the bleeder/breather sequence. The compaction pressure provided by a vacuum bag is crucial for good quality laminates when they are cured within a vacuum bag only. In an autoclave however, much higher pressures are available (>10 Bar/145 PSIG). The vacuum bag is still a necessary part of the process in the autoclave, because it acts like a membrane that separates the vessel pressure from the laminate. This is much like having an intimately conformable internal tool surface on the backside of the laminate in a large press. The resultant product is uniformly compacted throughout the vacuum bag.

Relating Vacuum and Atmospheric Pressure

The planet Earth is surrounded by gaseous matter that forms our atmosphere. This extends over 62 miles/100 km above the earth and is held to the earth by gravity. Being a gas, the atmosphere has weight, and that weight is normally measured in pounds per square inch (psi) or kilograms per square centimeter (kg/cm^2).

If you were to take a square inch column of the air extending to the outer atmosphere, its standard recognized weight and pressure exerted on the earth at sea level would be 14.7 lbs. This is called atmospheric pressure. Pressure measured that is greater than atmospheric pressure is referred to as gauge pressure (hence the use of PSIG units when describing autoclaves). Pressures below this are referred to as vacuum and are measured in (theoretically displaced) inches or mm of mercury (Hg).

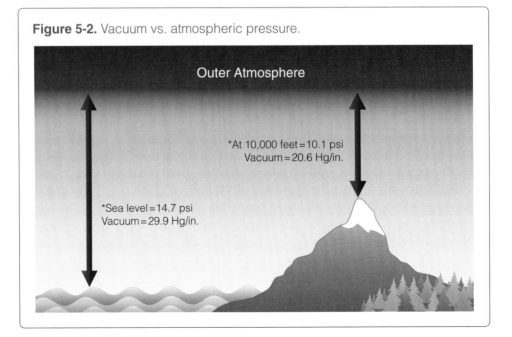

Figure 5-2. Vacuum vs. atmospheric pressure.

Outer Atmosphere

*At 10,000 feet = 10.1 psi
Vacuum = 20.6 Hg/in.

*Sea level = 14.7 psi
Vacuum = 29.9 Hg/in.

This same square inch column of air weighing 14.7 psi is equal to a square inch column of mercury (Hg) 29.92 inches high. (In other words, 29.92 in.^{3} of mercury would weigh 14.7 lbs on a scale.)

Atmospheric pressure decreases at higher elevations. The average loss is approximately 1 in/Hg (.46 psi) per 1,000 feet, up to about 13,500 feet. Beyond that, the atmospheric pressure decreases by 50% at an altitude of about 3.5 miles/5.6 km. This pressure drop is roughly exponential, so each doubling in altitude after this point results in a decrease of about half the measured pressure up to about 62 miles/100 km.

Vacuum Bagging Requirements

To obtain a good vacuum, certain requirements must be met:

- The vacuum bag and the base plate, tooling or mold must *not* be porous (i.e., must be airtight and hold a vacuum).
- Leaks must be eliminated, as much as possible.
- The vacuum pump must be large enough to accommodate the volume of the bagged area.
- The vacuum hose drawing air out of the vacuum bag must be large enough to provide necessary airflow and its length to the pump or storage tank should be minimized to reduce volume.
- A vacuum gauge should be used on to check that the required vacuum is achieved, and that there are no leaks in the bag.
- Caution should be used so as not to release the vacuum when installing a gauge, especially when using liquid resins. This can result in reintroduction of air into the laminate with no route to extract the air as the bleeder may already be saturated.

The breather and bleeder layers are another important factor in achieving good vacuum. The basic rules are:

1. There should be a continuous breather layer across the area being vacuum bagged. Without a breather the pressure differential is non existent resulting in a static pressure condition.

2. The breather should not be allowed to fill up with resin.

3. The breather and bleeder layers should make contact at the edges with a separator layer between to prevent saturation of the breather layer.

If both the bleeder and breather layers fill up with resin, the dynamics beneath the vacuum bag change from atmospheric pressure to hydraulic pressure (from the resin). When this happens, it is no longer possible to maintain an atmosphere of (down force) pressure across the bagged area.

For this reason, the aforementioned separator layer is often used between the bleeder and breather layers so that excess resin being pulled into the bleeder layer does not move up and saturate the breather layer. The bleeder and breather layers must make contact around the edges or through small, widely-spaced perforations so that air and volatiles have a continuous path to be extracted out of the laminate being cured.

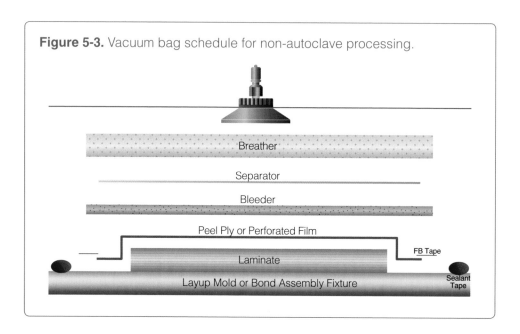

Figure 5-3. Vacuum bag schedule for non-autoclave processing.

Breather

Separator

Bleeder

Peel Ply or Perforated Film

FB Tape

Laminate

Layup Mold or Bond Assembly Fixture

Sealant Tape

Vacuum Bag Schedule and Function

Bleeder, Breather, and Vacuum Bag Schedule

The vacuum bag, bleeder, and breather schedule is made up of a series of different materials that are applied to the uncured composite laminate prior to processing. The purpose of using a vacuum bag is to draw out the excess resin, air, volatiles, and solvents from the laminate, while at the same time consolidating the layers into a tightly compacted laminate for the final cure cycle.

A bleeder layer is necessary in a non-autoclave process to promote movement of resin, and thus movement of air and gas. It may not be required in an autoclave process where several atmospheres of pressure are applied to the vessel/laminate.

Release Film or Peel Ply

The first layer that goes against the uncured laminate is typically a release film or a peel ply that is used as a barrier between the laminate and the subsequent bleeder or breather layers. This layer can be nonporous or porous material depending on whether or not resin bleed is necessary. Often a perforated release film is used for a controlled resin bleed. The diameter and the spacing of the holes can vary depending on the amount of resin flow desired. A porous peel ply is used when you do not wish to restrict the resin bleed and/or a peel ply surface texture is required. A nonporous peel ply commonly known as TFNP, is a fluorinated ethylene propylene (FEP), or Teflon® coated fabric that is used when no resin bleed is required, but evacuation of the volatiles and solvents is desired. This layer usually extends beyond the edge of the layup and can be sealed and/or secured with flash breaker (FB) tape as required.

Bleeder Layer

The bleeder layer is used to absorb resin from the laminate either through a porous peel ply or a perforated release film as described above. The bleeder layer is usually a non-woven synthetic fiber material that comes in a variety of different thicknesses and/or weights that range from between 2 oz./yd^2 to 20 oz./yd^2. Multiple layers can also be utilized for heavy resin bleed requirements. This layer usually extends beyond the edge of the layup and is secured in place with FB tape as required.

Separator or Blocker Film Layer

The separator layer is used between the bleeder layer and the subsequent breather layer to restrict or prevent resin flow. This is usually a solid or perforated release film that extends to the edge of the layup, but stops slightly inside the edge of the bleeder layer, to allow a gas path to the vacuum ports. Nonporous FEP can also be used as a separator layer.

Breather Layer

The breather layer is used to maintain a breather path throughout the bag to the vacuum source, so that air and volatiles can escape, and so continuous pressure can be applied to the laminate. Typically synthetic fiber materials and/or heavy fiberglass fabric is used for this purpose. The breather layer usually extends past the edges of the layup so that the edge band makes contact with the bleeder ply around the separator film. The vacuum ports are connected to the breather layer either directly or with strips that run up into the pleats of the bag. It is especially important that adequate breather material be used in the autoclave at pressure.

Bag Film and Sealant Tape

The bag film is used as the vacuum membrane that is sealed at the edges to either the mold surface or to itself if an envelope bag is used. A rubberized sealant tape or putty is used to provide the seal at the periphery. The bag film layer is generally much larger than the area being bagged as extra material is required to form pleats at all of the inside corners and about the periphery of the bag as required to prevent **bridging**. Bag films are made of Nylon®, Kapton®, or P.V.A. (polyvinyl alcohol) materials.

Vacuum Port

A machined or cast metal fitting that connects the vacuum bag to the vacuum source. The vacuum source can be a vacuum pump or a compressed air venturi. The port is connected to the vacuum source with a reinforced hose. Both the hoses and ports typically incorporate quick-disconnect fittings that allow the hoses to be removed from the ports without losing the vacuum on the bag.

Hand Layup – Wet Layup

Wet layup is the one of the simplest, least expensive, yet most labor intensive methods of composite manufacturing because it involves manually wetting out the reinforcement fibers with resin.

The fibers can be any of the normal reinforcements and forms available, however the resins are normally (but not always) cured at room temperature. These include polyester, vinyl ester and epoxy resin systems.

Wet layup is achieved by first coating the mold with a layer of filled resin often called a surface coat and/or a wet layer of laminating resin. Next, dry fiber reinforcement is placed onto the wet surface. More resin and fiber is added successively, and the resin is systematically worked into the fibers using a roller or a squeegee. Typically, a low viscosity resin will more easily wet-out the fiber but may run down vertical surfaces. A filled resin will hang better on the vertical surfaces but may take more effort to wet out the fibers. (*See* Figure 5-4.)

Wet layup processing is often performed with fairly unskilled labor; however, variation in skill level and handling techniques can result in significantly different laminate properties. For example, one worker may easily wet out the reinforcement using 1 oz/yd^2 of resin while another worker uses 4 oz/yd^2 of resin on the same amount of fabric. Consistent resin content is not easy to achieve using this method. Also, too much rolling to distribute the resin can actually break fibers and bring air back into the laminate causing porosity and voids. It is sometimes difficult to control the fiber orientation in a wet layup, as the materials have a tendency to slide around. This also decreases the laminate strength and overall performance.

Figure 5-4.
Hand layup – wet layup.

The temperature in the shop has a significant effect on the viscosity of the resin and its ability to wet out the fibers. Decreased temperature usually means more time to work the resin but perhaps a more viscous resin, making it harder to wet out the fibers. Increased temperature usually means decreased viscosity but also a reduced time in which to work the resin before it gels. Improvements can be made to the process by controlling the ambient temperature in the work space and prescribing the amount of resin that is used to wet out the materials.

Vacuum bagging a wet layup can enhance the quality of the end product. A properly designed bleeder and breather schedule allows for good compaction, excess resin extraction, and removal of trapped air and gas in the laminate.

Vacuum Infusion

Vacuum infusion processing is a cleaner and more precise process than wet layup because all reinforcements are placed dry, vacuum bagged and then resin is introduced through tubing using vacuum pressure and cured. (This is also called resin infusion.) Thus, workers never have to roll the resin into the reinforcements. The dry fiber reinforcement or "preform" is positioned in the mold (either with or without a surface coat layer), then a series of resin supply tubes/channels and vacuum plumbing is placed on the backside of the layup. A vacuum bag is installed over the entire assembly. (*See* Figure 5-5.)

Vacuum is pulled on the bag and the preform is compacted in place before resin is introduced into the laminate. The process works based on standard fluid dynamics, where the flow of the resin is dictated by the low-pressure differential created under the bag. The resin will flow until a pressure equilibrium is obtained. The distance that the resin will flow is determined by the amount of pressure available (amount of vacuum), the viscosity of the resin, and the permeability of the fiber preform. Multiple resin inlets are usually

Figure 5-5. Vacuum infusion (resin infusion).

molding methods and practices

97

installed at select locations to allow for infusion of larger sized parts. As a rule of thumb, no more than 30 inches/76.2 cm should be attempted between inlets with full vacuum infusion capabilities.

Vacuum infusion processing provides fairly consistent laminate properties with fiber volumes up to 70%. Because there are no reactive materials present until the resin is mixed, time for preform assembly and installation into the mold is not limited as it is with open mold wet layup. This allows for more precise control of material placement and fiber orientation. The amount of waste with this process is also minimized and volatile organic compounds (VOC) can be better contained.

As with wet layup, there are significant issues that have to do with the control of temperature, both in the shop and at the mold surface. A cold mold can greatly affect the viscosity of an epoxy resin, lessening the distance that the resin will flow. On the other hand, too much heat may cause the resin to gel before sufficiently filling the preform. There is also a significant risk of out-of-control exotherm with resin contained in the supply bucket or barrel, as this is usually a sizeable volume. These problems must be closely monitored during and after an infusion process.

Another issue with vacuum infusion processing has to do with the pre-compaction of the fibers. At the faying surfaces, the fiber to fiber interfaces are so tightly compacted that it is difficult to get resin wetting between fibers. As a result, the interlaminar shear strength and edgewise compressive properties of the laminate are somewhat compromised. For this reason, vacuum infusion processing may be limited to certain types of structures.

Another consideration with infusion processing has to do with achieving symmetry with the various forms chosen for this process. It is not unusual to select a multiaxial stitched form that does not produce a symmetric laminate even when it is flipped over at the mid-plane of the laminate. Caution should be used when designing laminates using these forms for infusion processing.

Hand Layup – Prepreg

Prepreg hand layup is one of the most widely used methods of fabricating aerospace composite parts and structures. Prepreg materials allow for very precise resin to fiber ratios and accurate control of fiber orientation in the final laminate. (*See* Figure 5-6.)

Prepregs are almost always processed within a vacuum bag, and cured in either an oven or an autoclave, depending on the required structural properties of the end structure. Although, some flat or gently shaped prepreg parts are hand laid-up and press-molded without the aid of a vacuum bag.

Prepreg layup is usually done in an environmentally controlled clean room where the temperature and humidity is regulated and monitored. While this is not required for some part manufacturers, it is commonplace in the aero-

Figure 5-6. Prepreg hand layup.

space industry. It makes sense to assign such strict controls to these materials as they tend to be fairly expensive and are used on many primary structural parts and airworthy aircraft components.

Since prepreg materials are already impregnated with resin, there is little to no mess involved with the material *kitting* (cutting to size) or the layup process. While manual kitting and layup is typical with prepregs, the materials also lend themselves to automated table cutting systems that are increasingly being utilized in high-rate production facilities. This, coupled with new generation laser positioning technologies for accurate material placement, makes prepreg layup one of the most ideal manufacturing methods available today.

While prepreg hand layup is perceived to be less labor intensive than wet layup, the reverse can be true when fabricating parts with complex geometries. In this case it may be necessary to layup, vacuum bag, and compact each ply individually, in sequence, in order to get the best results. These intermittent vacuum *debulks* are part of the normal prepreg layup process. Without these steps, *bridging* in the inside corners and wrinkling of the fibers on the outside corners may occur.

The tooling (molds) used for prepreg manufacturing is normally of better quality, and is typically made from better materials, than those used for wet layup processes. This is primarily because they are used in an elevated temperature environment and must maintain vacuum and/or pressure integrity through multiple thermal cycles. Tools of this type can be more costly up front and fairly expensive to maintain.

TABLE 5.2 Comparison of Wet Layup vs. Infusion vs. Prepreg

Wet Layup	Vacuum Infusion	Prepreg Layup
Advantages		
• Inexpensive	• Inexpensive	• Little mess or smell.
• Usually low-temperature.	• Usually low-temperature cure but elevated temp. cures possible.	• Lower exposure to health risks.
• Special equipment needs minimal.	• Less messy than wet layup.	• Easy to control resin/ fiber ratio accurately.
• Vacuum bag processing is an option but not always necessary.	• Containment of volatile organic compounds (VOCs).	• Many high-performance matrix systems available.
	• Easy to control resin/ fiber ratio accurately.	• Easier to cut to shape than many dry materials.
		• Long working time with most production prepregs.
Disadvantages		
• Messy	• Low wetting of fibers between layers.	• More expensive than wet layup or infusion.
• Can be smelly, especially with polyester and vinyl ester resins.	• Poor cosmetic surfaces without surface coat resin.	• Frozen storage required for most systems.
• Greater exposure to health risks.	• Runaway exotherm possibility in source buckets.	• Proper thawing technique critical to quality of finished part.
• Runaway exotherm is a possibility.	• Accurate cutting and placement of preforms is vital.	• Requires vacuum bagging and elevated-temperature curing, often requires autoclave cures.
• Limited amount of time to finish the lay-up.	• Special materials & equipment required.	• Sensitive to moisture, dirt, and oil contamination.
• Difficult to control resin/ fiber ratio accurately.	• Vacuum bagging is necessary.	

Oven and Autoclave Equipment

For advanced composite parts that require an elevated temperature cure, an oven or an autoclave (a cylindrical pressure vessel with heating and cooling capabilities) may be prescribed. The parts are generally laid-up, vacuum bagged, and placed in the oven or autoclave for processing.

Both ovens and autoclaves are normally fitted with vacuum plumbing and thermocouple connections. Autoclaves typically have advanced vacuum/ venting capabilities to allow the vacuum lines to be vented or regulated during the cure process. In addition, there may be special tracks or racks for loading and unloading large parts or groups of smaller parts.

Temperature controllers range from single-setpoint manual devices to full-blown computer-aided control systems that may be necessary to control hundreds of sensors in the oven or autoclave, and on multiple tools and parts during the curing process. Both ovens and autoclaves have heaters and fans that provide convection heating in a manner that uniformly distributes the heat throughout the vessel.

Figure 5-7. Production autoclaves.

Multi-vessel control panel with large production autoclaves in background *(left)*, and small-diameter production vessel *(right)*. (Photos courtesy American Autoclave Co.)

While ovens are normally rectangular in shape and usually have simple hinge/latch mechanisms (because they have no internal pressurization capabilities). Autoclaves are cylindrical in shape and have dome-shaped doors that are generally hydraulically actuated, with a rotating lock-ring or door-locking design to provide clamping, sealing, and safety functions during pressurization cycles. (*See* Figure 5-7.)

Autoclaves can be pressurized with air or an inert gas to many atmospheres of pressure (>10 atm or 147 psi) inside the vessel, providing much more compaction force on a composite laminate during the curing process. This allows for excellent consolidation of the materials within the vacuum bag, which intimately conforms to the inner laminate surface in this process.

Filament Winding

Filament winding is one of the oldest composite manufacturing methods and was originally invented to produce missile casings, nose cones and missile fuselage structures. The process is well-suited for automation and remains one of the most effective for mass production, delivering low production costs. Filament winding is now used to make a wide variety of structures, including pipes, golf club shafts, compressed natural gas (CNG) cylinders, composite oars, chemical storage tanks, power/transmission poles and fire fighter's breathing tanks. Basic filament winding is restricted to surfaces of revolution, but elaborate, 7-axis CNC-controlled winding machines can produce complex structures. (*See* Figure 5-8 on the next page.)

Filament winding involves feeding continuous filaments (fibers or rovings) through a resin bath and winding them onto a rotating **mandrel**. The fibers are drawn from a **creel** and through a collector that combines them into a tape. The mandrel rotates as the collector traverses its length. The combination of transverse movement and rotation creates a fiber angle on the structure. This angle can be changed by varying the speed of rotation and the rate of collector movement. The end result creates angles ranging from a hoop

Figure 5-8.
Filament winding.

IBK-Fibertec in Bueren, Germany uses a 1.6-m diameter, 22m long filament winding machine to produce large FRP storage tanks according to customer specifications. (Photo courtesy of IBK-Fibertec GmbH)

Figure 5-9. Variety of filament winding machines.

Filament winding machines are typically customized to meet specific project requirements, resulting in a wide variety of equipment possibilities, including machines with multiple spindles, like the machine shown on the left which can produce three structures simultaneously, and also very large machines, like that shown on the right which traverses the length of this wind blade root section by means of a carriage on top of steel rails. (Photos courtesy of Entec Composites Machines, a Zoltek company)

wrap, which is good for crush strength, to a helical wrap, which provides greater tensile strength. Successive layers may be oriented differently from the previous layer. Tension may also be precisely controlled, with higher tension producing a more rigid and higher strength end-product, and lower tension producing one with more flexibility. Filament winding machines can now be controlled using computer numerical control (CNC) technology, managing from 2 to 6 different axes of motion.

Once filament winding layup over the mandrel is complete, the composite structure is cured at room temperature or else the mandrel is placed in an oven for higher temperature cure. After cure, the mandrel is typically removed for re-use, leaving a hollow final product.

Filament winding excels at creating structures with very high fiber volume fractions (60–80%) and with specific, controlled fiber orientation. It provides one of the greatest strength to weight ratios of all composite manufacturing processes because of the continuous fibers. Glass fibers are the most common reinforcement used; however, carbon and aramid fibers may also be used. Resins used include epoxy, polyester, vinyl ester and phenolic. Thermoplastic polymers have also been used in the form of commingled fiber products and semi-preg tapes.

Automated Tape Laying and Automated Fiber Placement

Automated tape laying (ATL) and automated fiber placement (AFP) are similar processes used to apply resin-impregnated continuous fiber to a mold-tool or mandrel. Both use a computer numerically controlled (CNC) placement head and both require a significant investment in equipment.

ATL is the older process, with equipment sales peaking in the late 1980s, while AFP equipment became the more popular choice in the 1990s. AFP is said to be a combination of filament winding and ATL. Due to renewed interest in tape laying and continued evolution in both types of processes and equipment, today the terms are often used side-by-side or interchangeably.

ATL excels at reducing fabrication costs for larger composite panels with flat, single-curved, or reasonably gentle compound curved shapes, while AFP is more efficient in producing smaller structures with complex geometry.

ATL uses slit unidirectional prepreg tape (3 to 12 inches wide) fed from a spool incorporated into the tape laying head while AFP uses up to 32 separately controlled preimpregnated tows or narrow tapes (0.125 to 0.5 inch wide) fed from one or more creels on or near the head.

The ATL tape laying head includes one or more spools of tape, a winder and winder guides, a heater, a compaction shoe, a position sensor and a tape cutter or slitter. AFP machine heads typically include a heater and compaction roller, as well as tow/tape placement guides and cutting mechanisms. For both processes, the head may be located on a gantry suspended above the tool that enables movement over the mold surface or it may be located on the end of a multi-axis articulating robotic armature that moves around the tool or mandrel.

For both ATL and AFP, the position of the head is computer controlled, enabling very precise placement of the prepreg tow or tape. The computer software used to control the machine is programmed using numerical data from part design and analysis. This input defines the lay down pattern for the part, comprised of multiple courses, with a course being one pass of material of any length at any angle. After the required number of plies have been applied, the part and tool/mandrel are vacuum-bagged and placed in an autoclave or oven for curing.

AFP is typically a faster process, with speeds commonly in the range of 1,000 to 2,200 inches per minute. It also has the ability for tows to be added, dropped and added again at any time during the lay down process. ATL may be used to lay up a large charge or blank which is then transferred to a CNC cutting cell where multiple parts are cut out of the laminate. These cut pieces or part kits may then be stacked and hot drape-formed into structures such as stringers, ribs, C-channels, and I-beams or they may be laid onto contoured tools and cured.

Advantages of both ATL and AFP include high layup rates for large parts, resulting in lower labor costs and fewer human errors. Because the layup process is automated, parts made with these processes are more uniform and consistent. Because the placement heads automatically cut material at the end of each pass, both processes cite large reductions in material waste.

Disadvantages involve the high initial overhead cost for the equipment, which typically can run between $2,000,000 and $6,000,000. However, the computer controlled precision of automated tape and fiber placement processes make for a high degree of accuracy and repeatability in manufacturing. Many machines have cameras mounted along the gantry or in the head to record the layup and detect defects in-process. This greatly enhances the quality assurance of these parts.

AFP has been used to build complex-shaped structures that traditionally required extensive hand layup, including ducts, nozzle cones, tapered casings, fan blades, spars, channels, cowls and fuselage sections (such as those made for the Raytheon Premier and the Boeing 787 Dreamliner). ATL has been used to manufacture parts made from flat laminates such as empennage structural parts (spars, ribs, stringers, C-channel, and I-beam stiffeners) on the Boeing 777, and the Airbus A330 and A340-500/600. ATL has also been used to produce contour tape laid parts including horizontal/vertical stabi-

Figure 5-10. Automated tape laying.

Automated tape layup of a contour skin for the Boeing 777 *(left)* and contour lower skin with stringers, rib shear ties, and spar caps integrated for Airbus A330/340 horizontal tailplane *(right)*. (Photos courtesy of MAG Cincinnati)

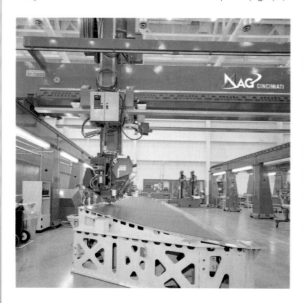

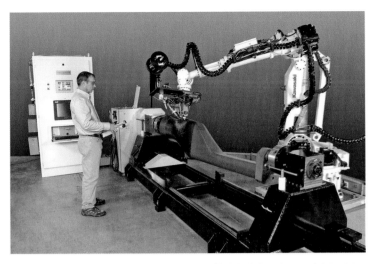

Figure 5-11. Automated fiber placement.

(Photos courtesy of Automated Dynamics Corporation)

The AFP/ATL-0510 Robot utilizes a Kawasaki Robot Arm platform for lab scale and production facilities. It hosts both fiber placement and tape laying heads for thermoset and thermoplastic processing.

The TP Simple Cylinder machine is a small, refurbished, existing gantry platform that was converted to a thermoplastic fiber placement machine for an industrial customer.

lizer skins for the Boeing 777 and A340-500/600, wing skins for the F-22 amd V-22 Osprey tilt rotor, B-1 and B-2 bomber wing and stabilizer skins, and Eurofighter upper and lower wing skins.

Materials

AFP and ATL have traditionally used carbon fiber prepregged with epoxy resin; however, more recent programs have used BMI and thermoplastic matrices, including PEEK, PPS, PEI, PA, PP, PE, PFA and others. The process is compatible with fiberglass, aramid, carbon fiber and other reinforcements.

Parameters

ATL delivery heads include a heating system that warms the prepreg tape to tack levels just ahead of the compaction shoe/roller. This is especially useful for BMI and other dry resin systems that lack sufficient tack for tape-to-tape adhesion. Heated tape temperatures range from 80°F to 110°F.

ATL is also said to provide "automatic debulking" via the compaction force typical delivery heads apply, which can range from 60 to 293 lbf across a 6-inch wide tape and as high as 601 lbf across 12-inch wide tapes.

AFP also claims the ability to tack materials together and debulk "on the fly" as the fiber placement head heats the materials at the nip point of the compaction roller, reducing the viscosity of the matrix resin and compacting the materials together simultaneously during lay down. In the case of thermoplastic composite placement, the materials are not only tacked together, but are bonded simultaneously. This eliminates the need for a post-process autoclave cycle.

Resin Transfer Molding

RTM is a process where mixed resin is injected into a closed mold containing fiber reinforcement. It produces a part with good surface finish on both sides and is well-suited to complex shapes and large surface areas. The procedure for RTM is to calculate the amount of fiber and core material desired in the final structure and place them in the mold without any resin. When the dry materials are prearranged in the correct sequence they are called a preform. The mold, which has been built to withstand vacuum and pressure, is then closed and resin is injected through one or more ports, depending on the size and complexity of the part.

Vacuum may also be applied and the mold may be heated to shorten cure-time. When the resin has cured, a molded part is removed for trimming and final finishing, and the process can be repeated.

RTM has traditionally been used for production of 5,000 to 15,000 parts per year. However, newer methods, such as the Floating Mold technology developed by VEC Technology, are enabling production volumes as high as 45,000 to 50,000 parts per year. Note that this process also uses composite tooling, which is much more economical than aluminum, steel and nickel-shell tooling, traditionally used for high-volume RTM production. In the VEC method, composite upper and lower mold skins are placed into a water bath and fastened to the outer walls of the universal vessel. The water bath fully supports both mold skins, eliminating the need for machined metal tooling.

RTM Materials

A wide range of resin systems can be used with RTM including polyester, vinyl ester, epoxy, phenolic, acrylic, methacrylates and BMI. Reinforcements may be continuous strand mat, unidirectionals, woven textiles and/or three-dimensional textile preforms.

Process Parameters

RTM is usually characterized as a low pressure process using pressures under 100 psi. However, higher pressures may be used, most commonly

Figure 5-12.
Resin transfer molding.

(Photos courtesy VEC Technology)

VEC Technology uses RTM to manufacture up to 45,000 parts per year. Here, a technician places kitted reinforcements into lower mold halves *(left)*.

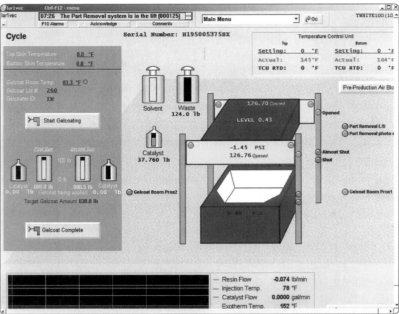

VEC's computer-controlled system (left) closes the mold and begins resin injection, which takes approximately 20 minutes *(below right)*.

After cure, the gantry-mounted upper mold halves are lifted and the molded parts are removed for trimming and attachment to other parts for final assembly.

Closed mold.

with matched metal tooling. RTM may use preheated resin and/or a preheated mold. A common example would be to use an epoxy resin preheated to 80°F and a composite or aluminum tool preheated to 120°F. However, high temperature variations have been successfully used to manufacture parts at temperatures close to 375°C (707°F) with long-term service temperatures in that same range. Process parameters are determined by the resin system used in conjunction with cycle time (viscosity and wet-out), desired T_g (service temperature) and post-cure requirements.

Pultrusion

Composite pultrusion is analogous to aluminum extrusion, but as the name suggests, fiber reinforcements are pulled (versus metals being pushed in extrusion) through a heated die to form a cured composite profile. Pultrusion can create almost any shape, including a wide variety of bars, beams, stringers, stiffeners, pipes and angles. These shapes can be solid or hollow, as long as the cross sectional area is constant. The heated die forms the shape without the use of conventional molds.

The pultrusion process starts with racks or creels holding rolls of fiber, mat or fabric reinforcement. As the reinforcement is pulled off the racks/creels, it is guided through a resin bath or resin impregnation system. Resin can also be injected directly into the die in some pultrusion systems. In the non-injected form of the process, reinforcements exit the impregnation area fully wet-out and are then guided through a series of custom tooling called a "pre-former," which helps arrange and organize the reinforcements into the final shape while excess resin is squeezed out.

The impregnated and shaped composite is now pulled into a heated die, which may have several different temperature zones to cure the composite matrix resin. Pulling in pultrusion may be achieved by using friction wheels or friction belts on either side of the profile die to pull on the shape as it exits the die. Another method uses reciprocating clamps that grip the shape and pull it in a continuous fashion in a "hand-over-hand" motion. Once the cured composite profile exits the die, it is then cut into lengths by a cutoff saw.

Figure 5-13. Typical pultruded composite products.

Small to medium sized pultrusions are shown here, illustrating the variety of crossections and shapes possible. This manufacturer can readily manufacture pultruded profiles up to 60 inches in width. (Photo courtesy of Sumerak Pultrusion Resource International)

Figure 5-14.
Typical pultrusion process.

(All images courtesy of Sumerak Pultrusion Resource International)

Continuous roving, mat and/or fabric materials travel from creels and through guides toward the resin impregnation and forming station.

The material infeed forms the fibers into the final shape before entering the pultrusion die.

Material exits the pultrusion die and heating system fully cured to final shape.

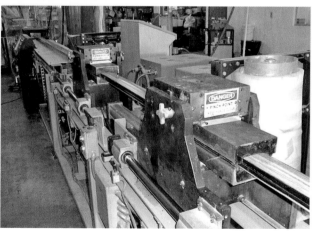

Dual reciprocating grippers pull the part at continuous speed. Next, a flying cutoff saw (not shown here) cuts the continuous profiles to final length.

More advanced pultrusion processes use a B-staged pultruded laminate, cutting it to length, and placing multiple pieces in a large multi-cavity die for processing. This is a cross between pultrusion and a matched-die forming process.

Pultrusion Materials

Thermoset resins are most commonly used, including polyester, vinyl ester, epoxy, phenolic and polyurethane resins. Thermoplastic polymers may also be used. The most common reinforcements used in pultrusion include fiberglass roving, continuous strand mat, woven fabrics and stitched fabrics. Other reinforcing materials may be used; however, more costly reinforcements, such as carbon and aramid fibers, tend to be employed in selected areas only, to achieve increased modulus and/or strength.

Process Parameters

Typically, pultrusion speeds are faster for thinner profiles. Profiles with a thickness of about 0.25–0.375 inch made with relatively fast-curing polyester resins can be pultruded at roughly 8 to 9 ft/min with fairly smooth surfaces. Thicker profiles ranging from 0.375 to 0.675 inch thickness can be made with a slightly slower-cure resin at speeds up to 6 ft/min with minimally abraded surfaces. Profiles over 0.67 inches in thickness made with medium reactivity resins can be produced at speeds of 0.5 to 5 ft/min and tend to have significantly more surface roughness.

Derakane quotes speeds as fast as 10 m/min (~33 ft/min) for profiles of 1 to 3 mm (0.04 to 0.1 inch) in diameter, and 1.5 to 1.9 m/min (~5 to 6 ft/min) for profiles between 4 and 9 mm (0.2 to 0.35 inch) in thickness.

Pultrusion is considered to be a low-pressure process, however, research has shown that pressures in the range of 3.5 to 14 bar (50 to 200 psi) can be developed with high filler and fiber contents.

Thermoforming

Note: Text from this section includes "The Thermoforming Process" from the April 2006 issue of *Composites Technology* magazine (copyright 2006, Gardner Publications, Inc.), used with permission.

Thermoforming is one of the oldest and simplest plastics forming processes. Baby rattles and teething rings were thermoformed in the 1890s, using cellulose-based plastics. The process saw major growth in the 1930s with the development of the first roll-fed thermoforming machines in Europe.

In the thermoforming process, heat and pressure are used to transform flat sheet thermoplastics (unreinforced or reinforced) into a desired three-dimensional shape. The sheet is preheated using one of three methods:

1. *Conduction* via contact heating panels or rods;

2. *Convection* heating, using ovens which circulate hot air;

3. *Radiant* heating achieved with infrared heaters.

The preheated sheet is then transferred to a temperature-controlled, pre-heated mold and conformed to its surface until cooled. The final part is trimmed from the sheet, and the trim can be reground, mixed with virgin material and reprocessed. This ability to recycle trim waste is typically most viable for unreinforced and some chopped glass-reinforced thermoplastics.

There are numerous variations of thermoforming, distinguished primarily by the method used to conform the sheet to the mold.

Sheet bending uses a folding machine or jig to produce a gently angled or bent surface.

Drape forming the preheated sheet is stretched down over typically a male mold using either gravity alone or application of vacuum.

Vacuum forming uses negative pressure (vacuum) to conform the material to the mold. *See* Figure 5-15.

Pressure forming uses positive pressure (compressed air or press equipment) to conform material to the mold.

Diaphragm forming is a lower pressure process suitable for simple geometries which uses air pressure to compress a flexible diaphragm against the sheet and conform it to the mold. *See* Figure 5-16.

Rubber forming, or rubber block stamping is a higher pressure process suitable for more complex geometries. It uses air pressure to compress an upper molding surface made from rubber to conform the sheet against the lower metal mold surface.

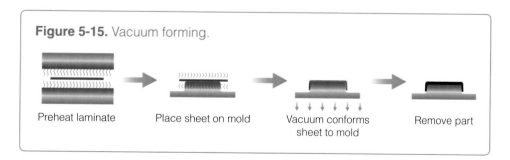

Figure 5-15. Vacuum forming.

Preheat laminate · Place sheet on mold · Vacuum conforms sheet to mold · Remove part

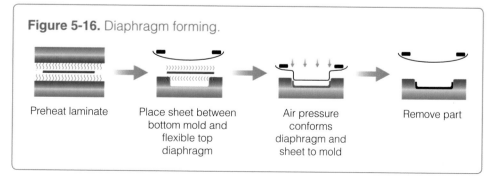

Figure 5-16. Diaphragm forming.

Preheat laminate · Place sheet between bottom mold and flexible top diaphragm · Air pressure conforms diaphragm and sheet to mold · Remove part

Hydroforming uses a pressurized fluid to conform the material. A thick rubber, flexible diaphragm, typically much larger than the part being formed, is attached to the upper platen of a press and supported from behind by a fluid medium, usually hydraulic fluid. The preheated composite material is placed on a lower mold surface, usually made from machined metal, which is attached to the lower platen. During compression, the fluid behind the diaphragm applies pressures as high as 69 MPa (~10,000 psi) for large systems, and produces more complex shapes in short cycle times. *See* Figure 5-17.

Matched die forming, or matched metal die stamping uses higher pressures and matched upper and lower metal molds. The mold is typically maintained below the thermoplastic melt temperature so that the part is formed and cooled at the same time. Cycle times range from 30 seconds to 15 minutes, with mold temperatures between room temperature and 350°F (25–177°C) and pressures from 100–1,500 psi. *See* Figure 5-18.

Twin-sheet forming positions two preheated sheets between two female molds with matching perimeters or contact surfaces. The twin sheets are drawn into the molds, using a combination of vacuum and air pressure, formed and then joined to make a single hollow part.

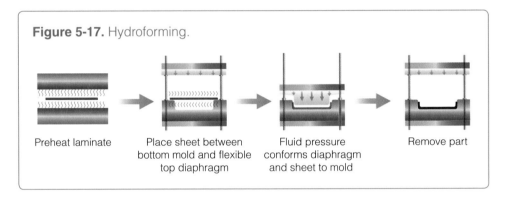

Figure 5-17. Hydroforming.

Preheat laminate — Place sheet between bottom mold and flexible top diaphragm — Fluid pressure conforms diaphragm and sheet to mold — Remove part

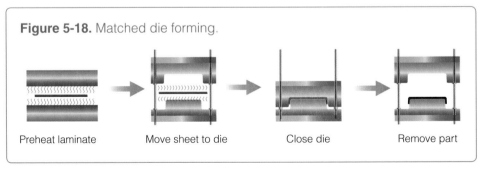

Figure 5-18. Matched die forming.

Preheat laminate — Move sheet to die — Close die — Remove part

Thermoforming Materials

Materials commonly used in thermoforming include commingled thermoplastic fibers such as Twintex (fiberglass/polypropylene made by Saint-Gobain Vetrotex) and the glass, carbon and aramid products made by Comfil using a wide variety of thermoplastic polymers including PET, LPET, PP, PPS, and PEEK.

There are also semi-pregs such as CETEX (made by Ten Cate Advanced Composites) and TEPEX (made by Bond-Laminates), which combine glass, aramid and/or carbon unidirectional or fabric reinforcements with thermoplastic matrices such as PEI, PPS, PP, PA (Nylon) and TPU. Porcher Industries offers similar combinations but in a range of powder-coated products.

Plytron is a unidirectional glass/PP prepreg made by Gurit. Cytec also makes a wide variety of thermoplastic prepregs.

There are also self-reinforced plastics (SRPs) such as Curv polypropylene fiber reinforced polypropylene (made by Propex Fabrics) and Milliken's Moldable Fabric Technology (MFT) sheet stock using PURE technology licensed from Lankhorst-Indutech.

Fiberforge uses an automated process to lay up multiple plies of continuous fiber reinforced thermoplastic prepreg into a flat, net-shaped preform called a Tailored Blank™. The fiber orientation in these blanks is tailored to meet the structural and other requirements of the final part. The tailored blank is consolidated into a composite laminate charge, which is then shaped into a 3-D part via thermoforming.

Figure 5-19. Thermoforming materials.

Reinforced thermoplastic materials for thermoforming include *(left)* Plytron unidirectional glass/polypropylene (PP) prepreg made by Gurit, and *(center)* CURV PP fiber reinforced PP made by Propex Fabrics, which is used for a variety of molded end-uses such as the part shown here from Samsonite's Cosmolite suitcase *(right)*. (Photos courtesy Gurit, and Propex Fabrics)

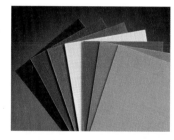

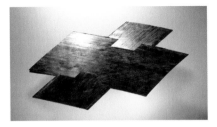

Figure 5-20. Thermoforming advanced composites from tailored blanks.

Fiberforge lays up multiple plies of continuous unidirectional fiber reinforced thermoplastic prepreg tape *(top left)* into a Tailored Blank™ *(bottom left)* made by an automated Relay™ Station, which is then consolidated before being thermoformed into a final composite structure like the carbon fiber/polyamide 6 automotive seat frame shown at right. (Photos courtesy of Fiberforge Corporation)

Thermoforming Parameters

Thermoforming temperatures depend on the specific thermoplastic being used, but typically range from 300–800°F (149–425°C). Forming pressures can range widely. Low-pressure processes may require as little as 15 psi for vacuum forming to over 1,500 psi for high pressure matched metal die forming. Typical pressures are around 500 psi.

For the Fiberforge Tailored Blank™ process described above, consolidation of the prepreg tape blank requires pressures of typically 220 psi and temperatures from 300–800°F (150–425°C), depending upon the thermoplastic polymer matrix used. After being consolidated, the Tailored Blank™ is then thermoformed using any of the sheet thermoforming methods described above. Typical processing in a matched mold takes 30–90 seconds using Fiberforge's 450-ton press.

For low-pressure thermoforming materials, tooling does not have to be steel but, depending on the length of a production run, can be aluminum or epoxy. Wood and even plaster tools can be used for prototyping.

The main difference between thermoforming and compression molding (see next section) is that the material is not expected to flow much in thermoforming. Instead, it is essentially forming a sheet that covers the full area of the part versus compression molding which involves putting a charge in a tool and expecting it to flow throughout the mold and fill all of its gaps. When higher pressures are used in thermoforming, it simply helps with surface finish, stamping fine detail into the parts, and forming certain complex geometries.

Compression Molding

Compression molding uses a heated matched mold similar to matched die forming, and has traditionally used thermoset sheet molding compound (SMC) or bulk molding compound (BMC) as molding materials. A hydraulic cylinder is used to bring the mold halves together, providing simultaneous heat and pressure. This causes the thermoset matrix to liquefy and flow into the voids in the mold where it chemically reacts and cures into the final shape. This molding process is capable of producing high volume, complex composite shapes while holding a very high tolerance. The material properties are less than that of conventional composite parts because of the discontinuous fibers in the molding material. Molds can be made from steel, cast iron or aluminum and are built with integral heating and cooling for rapid turnaround.

Compression Molding Materials

Originally, SMC and BMC materials were mostly made from polyester resin reinforced with chopped glass fiber. However, there are manufacturers of epoxy, cyanate ester, and BMI resin based bulk molding compounds used for manufacturing highly detailed structures with thin and thick sections, as well as molded-in features such as threads and inserts for satellite, aerospace and defense, commercial aircraft, industrial and sporting applications.

No longer confined to thermoset materials, compression molding is used to mold glass mat thermoplastic (GMT) and long fiber reinforced thermoplastic (LFRT) materials into automotive hoods, fenders, scoops and spoilers. There are a wide variety of glass mat thermoplastic (GMT) thermoformable sheet materials made by companies such as Azdel, Quadrant Plastic Composites, Owens Corning and FlexForm Technologies (which uses natural fibers such

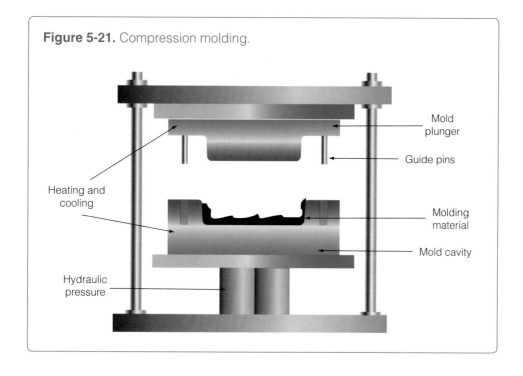

Figure 5-21. Compression molding.

as kenaf, hemp, flax, jute or sisal, blended with thermoplastic matrix materials, such as polypropylene or polyester).

Advanced thermoplastic composites can also be compression molded using unidirectional tapes, woven fabrics, randomly orientated fiber mat or chopped strand products. Fiberglass and carbon fiber are the most common reinforcements.

Process Parameters

SMC is typically molded at pressures close to 1,000 psi, and sheet materials competitive with GMT can require pressures as high as 3,000 psi. There are also low-pressure SMC materials which only require 150 psi. Thus, these low pressure molding compounds (LPMCs) require only a 700- to 750-ton press versus the 3,000-ton presses more common with traditional SMC and BMC materials.

Core Materials

6

Why Use Sandwich Construction?

A sandwich structure consists of two relatively thin, stiff and strong faces separated by a relatively thick, lightweight core material. The face sheets are usually connected to the core material with an adhesive. *See* Figure 6-1.

Sandwich construction is used to achieve structures with high strength and high stiffness that are still lightweight. A sandwich structure acts like an I-beam, which can withstand great loads without bending (high stiffness) or failing (high strength). The faces of the sandwich panel act as flanges and the core acts like the I-beam's web, connecting the load-bearing skins. Depending on the type and properties of core used, it is possible to dramatically increase the stiffness and strength of a structure without increasing its weight. As shown in Figure 6-2, this increase is directly related to the thickness of the core.

Other reasons for using core may include additional benefits offered by specific core materials:

- Thermal insulation or thermal transfer.
- Dampening of vibration and noise.
- Filling of hollow spaces to limit water ingression.

Sandwich vs. Solid Laminate

Sandwich construction is just one type of structural composite design. Although cored construction has many benefits, it also has some disadvantages, which may dictate alternate construction for certain applications. Solid laminates offer better resistance to damage and better damage tolerance than do cored structures; but they also tend to be heavier. Solid laminates are very effective where the overwhelming need is for strength versus stiffness, in which case a thin solid laminate is more opti-

In this chapter:
- Why Use Sandwich Construction?
- Balsa Core
- Foam Cores
- Honeycomb Cores
- Other Core Types
- Design and Analysis
- Fabrication
- In-Service Use

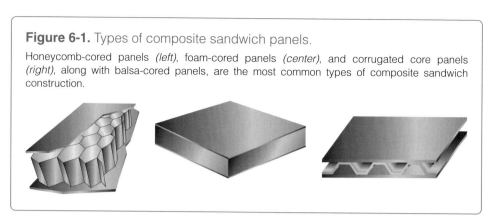

Figure 6-1. Types of composite sandwich panels.

Honeycomb-cored panels *(left)*, foam-cored panels *(center)*, and corrugated core panels *(right)*, along with balsa-cored panels, are the most common types of composite sandwich construction.

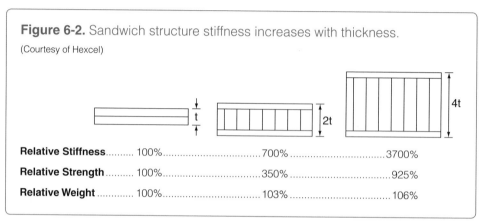

Figure 6-2. Sandwich structure stiffness increases with thickness.

(Courtesy of Hexcel)

Relative Stiffness.......... 100%................................700%.................................3700%		
Relative Strength .... 100%................................350%...................................925%		
Relative Weight .... 100%................................103%...................................106%		

mized to meet the high-strength loading demands using the least amount of material.

Skin-stringer, hat-stiffened or rib-stiffened structure can be described as a plate that is reinforced periodically by hollow, solid or cored beams. It is the most common alternative to sandwich construction in aircraft structures. It is also common in non-cored boat hulls.

- Stringers are solid laminate reinforcing beams, usually in the form of I-beams or J-sections.
- Hat stiffeners are hollow or foam-cored beams having a rectangular or trapezoidal cross-section.
- Ribs are thin, solid plates or verticals used as reinforcing members.

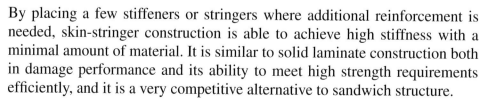

By placing a few stiffeners or stringers where additional reinforcement is needed, skin-stringer construction is able to achieve high stiffness with a minimal amount of material. It is similar to solid laminate construction both in damage performance and its ability to meet high strength requirements efficiently, and it is a very competitive alternative to sandwich structure.

Corrugated construction offers stiffness approaching that of sandwich construction combined with the damage tolerance of a solid laminate and perhaps the easiest and most cost-effective fabrication, especially in wide spans.

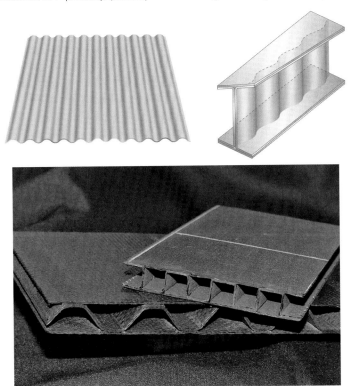

Figure 6-3. Corrugated composite construction.

Corrugated composite sheet *(left)*. Sine-wave spar *(right)*. FyreRoc fireproof composite panels made by Goodrich EPP *(bottom)*. (Courtesy of Goodrich Engineered Polymer Products)

It is well-suited to continuous processing. Common examples of corrugated construction include:

- Corrugated fiberglass sheets used in greenhouses, skylights, siding and roofs.
- "Sinusoidal stringers" or "sine-wave beams" used in aircraft spars and other stiffness-driven structures.
- Corrugated composite laminates used as a sandwich core in composite bridge decks and other heavily-loaded structures.

Balsa Core

Balsa wood is the most commonly used wood core in composite sandwich structures. Balsa is very light yet exceptionally strong. Although classified as a hardwood, its density ranges from only 4 to 20 pounds per cubic foot (pcf). Balsa gets its strength from its vascular system. Similar to honeycomb, this vascular system is capable of carrying tremendous compressive and shear stresses.

End-grain balsa is produced by cutting harvested and kiln-dried balsa wood across the grain. These cut pieces are assembled into blocks and cut into sheets, or cut into smaller blocks which are then glued onto a scrim fabric to produce a drapeable sheet. The grain orientation of the final balsa core is in the z-direction, or perpendicular to the face-skins, which gives the maximum strength and stiffness.

Balsa continues to be used in many cored composite applications due to its high stiffness, compressive properties and low cost. Although it was used in early applications of aircraft flooring, it is no longer common in aircraft due to increased stringency of smoke and flammability regulations as well as a general trend toward lighter core materials such as honeycombs and foam.

Balsa and foam core materials are typically used in boat construction, with foams traditionally being used more than balsa in Scandinavia and Europe. Balsa remains popular in cored boat hulls and decks, especially as resin infusion becomes an increasingly popular fabrication method. Open-celled materials such as honeycomb are not easily compatible with resin infusion, because the cells fill up with resin. Honeycomb and foam are also more expensive than balsa.

Figure 6-4.
This illustration shows the vascular system of balsa wood, similar in structure to honeycomb, which enables it to withstand large compressive and shear stresses.
(Courtesy of I-Core Composites)

Figure 6-5. Balsa blocks and end-grain balsa sheets.
End-grain balsa core is made by cutting balsa across the grain, assembling these cut pieces into blocks *(left)*, and slicing these blocks into sheets *(right)*; these may then be further processed into contourable and infusible panels. (Courtesy of I-Core Composites)

Foam Cores

Foam is made by adding foaming or blowing agents to polymer materials (e.g., resins or melted plastics), which causes gas pockets to form within the structure of the polymer as it cures or hardens. Foaming agents may be chemical or physical.

- Physical foaming agents include pressurized gases, such as nitrogen.
- Chemical foaming agents are materials that cause a chemical reaction which produces gas.

Most cores used in composite sandwich structures are rigid and closed-cell, meaning that each gas cell is completely surrounded by cured resin and thus isolated from all other cells. A closed-cell structure prevents water migration through the core material, which assists in minimizing moisture ingress and absorption.

Foams are manufactured from a wide variety of polymers and can be supplied in densities from less than 2 pounds per cubic foot (pcf) to 60 pcf.

TABLE 6.1 Foam Core Materials

Polymer	Trade Name	Manufacturer	General Applications
Polyvinyl Chloride (PVC)	Airex	Alcan Airex	• Composite sandwich
	Divinycell	DIAB Group	
	Klegecell	DIAB Group	
Polyurethane (PUR)	Last-A-Foam	General Plastics	• Composite sandwich
	Polypro	Polyfoam Products	• Cushioning
	Stepanfoam	Stepan Company	• Packaging
	Polyumac	Polyumac	• Buoyancy/flotation
		Mgi	• SIPs
			• Insulation
	Modipur	Hexcel	• Models
Polyisocyanurate (ISO)	Elfoam	Elliott Company	• Insulation • Composite sandwich
Polystyrene (EPS or PS)	Styrofoam	Dow Chemical	• SIPs
	EPS	Benchmark Foam	• Refrigeration • Insulation
Polyetherimide (PEI)	Airex R82	Baltek	• Structural cryogenic insulation and high fire resistance
Polymethacrylimide (PMI)	Rohacell	Evonik	• Composite sandwich
Co-polymer **SAN**	Corecell	Gurit	• Composite sandwich
Hybrid	Divinycell HD	DIAB Group	• Composite sandwich
	Renicell	DIAB Group	
	Ultracore	Futura Coatings	
Syntactic Foam **Polyester**	Spray-Core	ITW SprayCore	• Print barrier
Epoxy	Scotch-Core Syntactic Film	3M	• Sub. for thin honeycomb
Multiple resins	Microply	YLA	• Sub. for faceskin plies
Cyanate Ester	BryteCor EX-1541	Bryte Tech.	• Composite sandwich
BMI	FM 475	Cytec	

Linear vs. Cross-linked PVC

Linear foams are made from thermoplastic polymers and have no cross-links in their molecular structure. (*See* Figure 6-6.) Thus, linear PVC foam is a purely thermoplastic foam, as opposed to cross-linked PVC which is a blend of thermoplastic PVC and thermoset polyurethane. The non-connected molecular structure of linear PVC foam results in higher **elongation** and **toughness**, but somewhat lower mechanical properties. It also enables significant deflection without failure, which means better impact resistance and energy absorption. Linear PVC foam is also easier to thermoform around curves; however, it is more expensive than cross-linked PVC foam, and has a lower resistance to elevated temperatures and styrene. Cell diameters range from .020 to .080 inches.

Cross-linked foams incorporate thermoset polymers and have cross-links between the molecular chains, producing higher strength and stiffness but less toughness. Elongation for cross-linked PVC foam is 10–20% versus 50–80% for linear PVC foam. Cross-linked PVC is more rigid—producing stiffer panels—but is also more brittle and prone to forming 45° cracks under impact. It is cheaper than linear PVC foam and less susceptible to softening or creeping at elevated temperatures. Cell diameters range from .0100 to .100 inches (compared to .0013 inches for balsa).

Polyurethane vs. Polyisocyanurate Foam

Polyurethane and polyisocyanurate foams are similar in that both are closed cell foams that have high R-values (typically between R-7 and R-8 per inch). These foams have higher R-values than most other foams because they trap small bubbles of gas during the foam manufacturing process. The gas is usually one of the HCFC or CFC gases, which have twice the R-value of air. Both kinds of foam can be sprayed, poured, formed into rigid boards and fabricated into laminated panels.

Polyiso foams have serveral advantages over polyurethanes, however:

- Improved dimensional stability over a greater range of temperatures.
- More fire-resistant.
- Slightly higher R-value.

Because of its low mechanical properties, polyisocyanurate foam is not really used as a structural core material. Although it is used as a foam core in composite sandwich structures, it is primarily an insulation layer due to its low

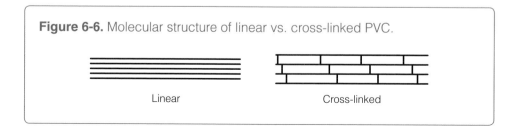

Figure 6-6. Molecular structure of linear vs. cross-linked PVC.

Linear   Cross-linked

density and high insulative properties. The number one use of polyisocyanurate foam is insulation in construction, where it is used in more than half of new commercial roofing applications and nearly 40% of residential sheathing applications. Polyiso foam is also used as insulation on the Space Shuttle Orbiter. One of the most easily recognizable parts of the Space Shuttle is the large, orange external fuel tank (the Space Shuttle uses two). The orange color is from a thin layer of polyisocyanurate foam covering the entire tank structure. The foam keeps the liquid oxygen (LOX) fuel inside the tanks from boiling. LOX is a cryogenic fuel with a boiling point of around -200°F.

Co-polymer and PMI Foams

Corecell is a linear, styreneacrylonitrile (SAN) co-polymer foam that was originally developed as a tougher alternative to PVC foam core for marine applications. Formerly manufactured by ATC Chemicals, Corecell is now part of the SP (Magog, Quebec) product line sold by the Gurit global composites group. Used as a structural core in composites, Corecell is available in densities from 3 to 9 pcf. It offers high shear elongation and impact strength, and is compatible with polyester, vinyl ester and epoxy resins. Applications include boat hulls, decks and other structures.

Corecell has become very popular for use with resin infusion processing. SP has completed extensive testing on over 30 different types of core cuts for infusion processing. Parts having curvature will need to use cut or scored foam core so that it can conform to the desired shape. However, with vacuum infusion processing, additional surface cuts or grooves are often added to aid in resin flow. The drawback is a significant weight gain with heavily cut foam, as well as a noticeable effect on mechanical properties and impact performance due to the additional resin cured in the numerous channels cut into the core. According to SP, making too many cuts introduces too much resin at one time. The ideal process is what SP calls "staged infusion," in which the appropriate number of cuts are used to impel the resin front in combination with multiple ports across the part, in order to re-introduce resin as the impregnating front passes through.

Rohacell® is a structural foam core material made from polymethacrylimide, or PMI. It is produced by mixing methacrylic acid and methacrylonitrile monomers, which react to form the PMI polymer. The monomer mixture, along with various additives are heated to a foaming temperature, which can be as high as 338°F depending on the desired density and grade of the foam. After foaming, the block is cooled to room temperature and the slick skins that form on the outer surfaces are removed. The block is then machined to produce foam sheets of various thicknesses.

The compacted, higher density Rohacell® foams have an oriented cell structure due to the manufacturing method used. Similar to end-grain balsa wood, higher density Rohacell® is cut into blocks with the cell orientation in the vertical direction and then bonded together. The resulting sandwich core has excellent properties in the thickness direction. Rohacell® is more water absorbent than other structural foam materials and it burns easily. However, it can withstand processing temperatures of up to 350°F. It is most commonly used in the aerospace industry in a variety of structural sandwich applications in commercial, military and general aviation aircraft, satellites, and helicopters.

Syntactic Foams

Syntactic foam core materials are made by mixing hollow microspheres of glass, epoxy and/or phenolic into fluid resin. Additives and curing agents are also combined to form a fluid which is moldable, and when cured forms a lightweight cellular solid.

Syntactic foam core can be produced from polyester, vinyl ester, phenolic and other resins, and is typically spray applied in thicknesses up to 3/8-inch and in densities between 30 and 43 pcf. Syntactic cores and non-woven materials (or laminate bulkers) are often used in the same manner to build up laminate thickness and increase flexural properties at minimal cost. They are not intended as structural core materials.

Honeycomb Cores

Honeycomb, foam and balsa wood are all low density, cellular materials. The cells in balsa and foam are really bubbles in the constituent material (wood or polymer resin, etc.). The cells in honeycomb are actually large spaces between formed sheets of paper, metal or other material. Thus, honeycomb is an "open-celled" material as the cells are really open spaces, where most foams and balsa are described as "closed-cell" because the cells are sealed bubbles.

Hexagonal is the most common cell shape for honeycomb cores, but others are also used including tubular, triangular/sinusoidal, and corrugated. Within hexagonal cell honeycomb, there are several additional shape variations available. These include over-expanded for conforming to simple or 2-D curvature, and specially-shaped "flex" cells for complex or 3-D curvature.

Over-expanded cell configuration is available from a variety of suppliers for Nomex® honeycomb—including M.C. Gill's OXHD honeycomb—and from Hexcel for its HRP impregnated fiberglass fabric honeycomb (sold as OX-Core®).

Flex-Core® is a patented technology developed by Hexcel. It enables easy forming of honeycomb into complex curvatures and shapes without loss of mechanical properties. It is offered for a number of Hexcel's honeycomb cores including those made from HRH-10 Nomex® paper, HRH-36 Kevlar® paper, HRP fiberglass cloth and for aluminum honeycombs made from 5052 or 5056 alloy foil. Double-Flex Core is only offered for Hexcel's aluminum

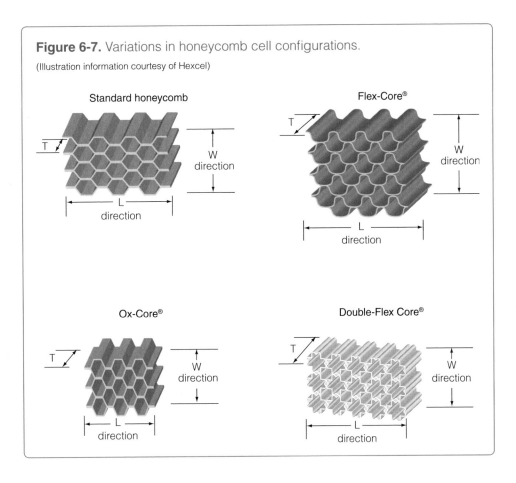

Figure 6-7. Variations in honeycomb cell configurations.

(Illustration information courtesy of Hexcel)

honeycombs. UltraFlex is a flexible honeycomb cell shape offered by Ultracor Inc., which can be made with any nonmetallic honeycomb and cell size. It uses a different cell configuration than Hexcel's Flex-Core® product and is also not formed via an expansion process.

L (Ribbon) vs. W (Transverse) Properties

Hexagonal honeycomb does not have the same properties in all directions. As a result of how hexagonal honeycomb is made, shear properties in the L or Ribbon direction are much higher than those in the W or Transverse direction. The Ribbon direction runs along the continuous sheets of web material which make up the honeycomb structure. The Transverse direction runs perpendicular to the ribbon direction. It is the direction that runs across the sheets of web material as opposed to along the sheets. Another way of easily defining the W direction is that it runs from flat side to flat side across the hexagonal cells.

How Honeycomb is Manufactured

Expansion (*see* Figure 6-8) is a process in which web material is fed into machines that apply ribbons of adhesive. (These ribbons of adhesive will form the *node-lines* for the expanded core material. Nodes are where two sheets of web material are bonded together with adhesive.) Sheets are then cut and stacked in layers, with every other sheet offset a specified amount, so that the resulting stack forms a honeycomb pattern when expanded. This HOBE (HOneycomb Before Expansion) block is expanded and honeycomb blocks made from paper are then dipped in resin (typically phenolic). Multiple dippings are sometimes required to produce desired cell wall thickness, honeycomb density and mechanical properties. Blocks are then cured. Horizontal slices are sawed from the honeycomb block to required thickness. Alternatively, slices of the desired thickness may be sawed from the HOBE and then expanded and dipped to produce desired properties.

There are two ways to increase honeycomb core density: increase sheet or web thickness and/or increase resin coatings. Extremely thick webs (> 4–5 mils thickness) are difficult to expand. For a given web thickness there is a limit to which additional dippings only increase weight and do not add significant properties.

Corrugation (see Figure 6-9) is typically used to produce higher density honeycomb cores, as well as sinusoidal and corrugated cores. Densities of 12 to 55 pcf are produced from this process using aluminum foil web material. For hexagonal honeycomb, web material is fed through corrugating rolls. The corrugated web material is then cut into sheets. Adhesive may be applied before or after corrugation. Sheets are stacked to form corrugated blocks. The honeycomb blocks are sliced into required thickness honeycomb sheets.

Figure 6-8. Expansion honeycomb manufacturing processes.
(Illustration information courtesy of Hexcel)

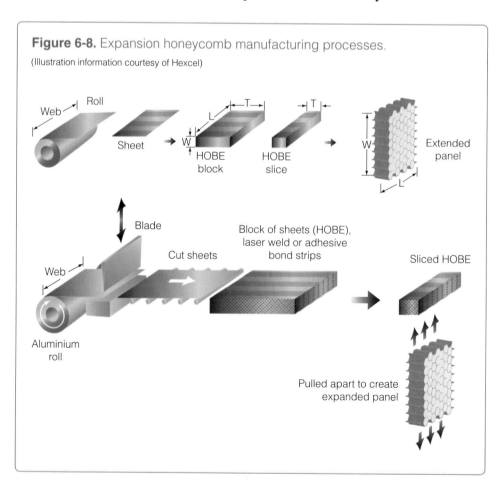

For sinusoidal honeycomb, two separate rolls of web material are used. One is fed through corrugating rolls, and the other is fed through a roll that applies adhesive. The two are adhered together by passing through a nip roll. This bonded sheet product is called *single face*. The single face now passes through slitting blades which produce multiple narrow bands of single face. These narrow bands are then laid side-by-side and squeezed together to form line boards. One outer surface of the line boards is coated with adhesive. The line boards are then stacked on a combining table to form large sheets of corrugated core. The thickness of the core is defined by the width of the slit bands of single face: wider bands produce thicker core, narrower bands produce thinner sheets of core.

Figure 6-9. Corrugation honeycomb manufacturing processes.

(Illustration information courtesy of Hexcel)

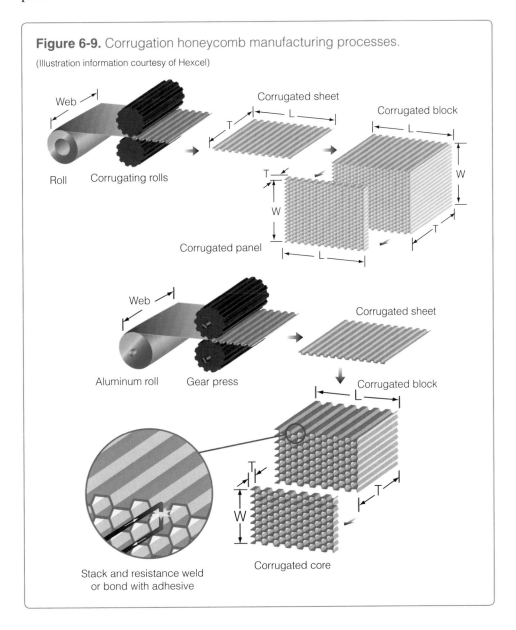

Figure 6-10. Sinusoidal honeycomb.

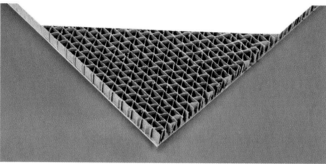

TRICEL honeycomb is made from kraft paper to form a continuous series of triangular cells with up to 95% open space and density ranging from 1 to 3 lb/ft^3. (Courtesy of Tricel Honeycomb Corporation)

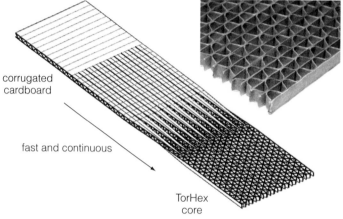

corrugated cardboard

fast and continuous

TorHex core

Patented TorHex honeycomb adds lengthwise slitting and in-line rotation of the strips to maximize large-scale production at high speeds. (Courtesy of EconCore—www.econcore.com)

Honeycomb Materials

Honeycomb can be made from a wide variety of materials, including: cellulose papers such as Kraft paper and cardboard; more advanced papers such as those made from Nomex® meta-aramid and Kevlar® para-aramid fibers; fabrics from fiberglass, aramid and carbon fibers; foils and metal sheet; and plastics. Carbon-carbon honeycomb is made by starting with epoxy-impregnated carbon fiber fabric, and heating it to 3,000°F. This carbonizes both the reinforcement and matrix resin, producing a carbonized fiber reinforced carbonized matrix composite honeycomb core.

TABLE 6.2 Honeycomb Core Materials

Material	Cell Shape	Product Name	Manufacturer
Cellulose Papers			
Kraft paper	Sinusoidal/corrugated	Tricel	Tricel Honeycomb
		Paper honeycomb	ATL Composites
Advanced Papers			
Nomex®	Hexagonal	ACH, APH	Aerocell
	OX	ACX, APX	Aerocell
	Hexagonal	ECA, ECA-I	Euro-composites
		HRH-10, HRH-78, HRH-310	Hexcel
		Gillcore HA, Gillcore HD	M.C. Gill
		PN1, PN2	Plascore
N36 Kevlar® Paper	Hexagonal	KCH	Aerocell
		HRH-36	Hexcel
		PK-2	Plascore
Fabrics			
Fiberglass	Hexagonal	HDC, HFT, HRP, HRH-327	Hexcel
Kevlar®	Hexagonal	HRH-49	Hexcel
		UKF	Ultracor
Carbon	Hexagonal	Triax*	Ultracor
Quartz	Hexagonal	UQF	Ultracor
Metals			
Aluminum	Hexagonal	ECM, ECM-P	Euro-composites
	Hexagonal	ACG, CR-III, CR-PAA	Hexcel
	Corrugated	Rigicell	Hexcel
	Hexagonal	Dura-Core II, PAA Core	M.C. Gill
	Corrugated	Higrid	M.C. Gill
	Hexagonal	PCGA-3003, PCGA-5056, PAMG-5002	Plascore
Stainless Steel	Hexagonal	SSH	Plascore
Plastics			
Polyurethane	Hexagonal	TPU	Hexcel
Polycarbonate	Tubular	PC2	Plascore
		PC Tubus Core	Tubus Bauer
Polypropylene	Tubular	PP Tubus Core	Tubus Bauer
	Hexagonal	Nida-Core	Nida-Core
Other			
Carbon-Carbon		PCC, UCC	Ultracor

* plus 50 other types of carbon fiber honeycomb.

Other Core Types

Note: Text from this section is adapted from "Engineered to Innovate" by Ginger Gardiner, *High Performance Composites Magazine*, September 2006; reproduced by permission of *High Performance Composites Magazine*, copyright 2006, Gardner Publications, Inc., Cincinnati, Ohio, USA.

Truss Cores and Reinforced Foam

TYCOR®

WebCore manufactures fiber-reinforced foam core materials, trademarked TYCOR®, using a proprietary manufacturing process in which fiber reinforcement is stitched through the thickness of closed-cell foam sheets to form a vertical web or rib-stiffened core structure. The dry fiber web and face skin fabric are processed using resin infusion or RTM to form a sandwich panel with z-directional reinforcement.

TYCOR® products include both a low density and a high density version, as well as custom products made to meet specific customer demands.

TYCOR® has been used in many bridge and infrastructure projects. It is also being evaluated by GE for an all-composite sandwich fan case for the newest GE90 and GEnx jet engines. The TYCOR® product being used here is custom-designed, with carbon fiber reinforcement and rigid 100% polymethacrylimide (PMI) closed cell foam that provides high performance at lighter weight than honeycomb. TYCOR® is also part of the only composite version among the final design solutions for the U.S. Army's redesigned AMX portable airfield mat system. PMI foam reinforced by a carbon non-

Figure 6-11. WebCore TYCOR® products.
(Courtesy of WebCore Technologies)

Product	Description
G6	Lightweight core for a wide range of applications. Replaces 5 to 6 lb/ft³ foams and balsa wood.
G18	Structural core for high strength applications. Replaces end-grain balsa and high density foams up to 15 lb/ft³.

crimp fabric with epoxy resin as a matrix are being used in 4-foot by 7-foot (1.2m by 2.1m) panels, with aluminum extrusion on four sides used for joining the panels together.

X-COR and K-COR

Albany Engineered Composites produces X-COR by reinforcing lightweight PMI foam with very thin pultruded carbon fiber/epoxy rods—0.28 to 0.5 mm/0.011 to 0.020 inches—arranged in a tetragonal truss network. These rods protrude from the foam, and are embedded into the face sheets during sandwich panel manufacture. The truss network carries both shear and compressive loads, while the foam base provides buckling support.

By changing the rod placement angle and insertion density, X-COR can be precisely engineered to meet unusual and specific loading patterns. In addition, high compression loads can be handled in precisely defined areas for hard point and fastener installation or for high point loading scenarios—without adding undesirable weight and labor cost associated with potting compound.

X-COR has been developed for co-cured structures with gentle contours or thick (>1.27 mm/0.050 inch) face sheets. It can be thermoformed and supplied as net-shape details and can be processed using the same cure techniques employed on honeycomb-cored structures. It is being used in a new all-composite tail cone for the U.S. Army's UH60M *Black Hawk* helicopter.

K-COR is a variant of X-COR, in which the 0.50 mm/0.020-inch diameter or greater reinforcing rods are set flush to the foam surface. Face sheets are either co-cured or secondarily bonded to the core. K-COR is thermoformable and better suited for more complex contours. Because the carbon rods used in K-COR have a more closely matched coefficient of thermal expansion carbon/epoxy face skins than unreinforced foam, this type of core can better withstand the extreme temperature differentials faced by space and missile structures. For this reason, as well as its excellent damage tolerance, K-COR is being qualified for use on the payload shroud of Boeing's *Delta IV* launch vehicles.

Figure 6-12. X-COR and K-COR.
Photo of X-COR *(left)* and model of K-COR *(right)*. (Courtesy of Albany Engineered Composites)

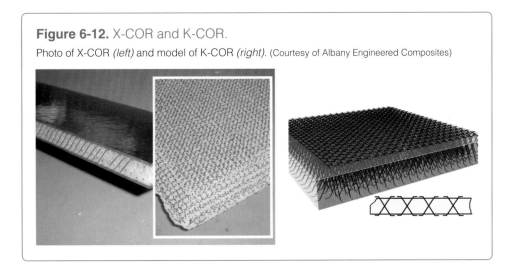

Non-Wovens

Bulking Materials vs. Print Barriers

The non-woven materials described here are generally referred to as bulking materials versus print barrier materials. The latter are also non-woven fabrics, typically produced by the same manufacturers as core fabrics or bulking materials; however, these materials are not used as cores. Instead, they are used as thin layers placed behind the gelcoat in boat lamination, to prevent the weave of the fabric reinforcement in a composite laminate from "printing" or showing through to the exterior of the gelcoat. Many of these materials are used in vacuum infusion processing to provide channels for resin distribution as well as bulk within the infused laminate.

Examples of applications are wide-ranging (*see* Table 6.3). Soric XF3 has replaced honeycomb in the dorsal cover on the U.S. Air Force *T-38 Talon* jet trainer, when the part was transitioned from autoclaved cured prepreg and honeycomb core to a vacuum infusion of dry reinforcements and non-woven core with vinyl ester resin. Spherecore has been successfully used in Olympic medal-winning composite kayaks and canoes. The Atvantage CSC and CIC materials are used in the interiors of private and business jet aircraft.

Figure 6-13. Non-woven core materials.

Soric XF3 combines a non-woven core and resin infusion flow medium into one material. (Courtesy of Lantor BV)

TABLE 6.3 Non-woven Core Materials

Tradename	Manufacturer	Form	Fiber
Lantor Coremat®	Lantor Distributed by Baltek	1 to 5 oz/yd^2 spun-bonded fabric	Polyester/thermoplastic micro-spheres/styrene-soluble binder
Spherecore **SBC**	Spheretex	Stitch-bonded web (resembles felt)	Fiberglass/ thermoplastic microbal-loons (note that carbon, aramid and other fibers can be used)
Soric	Lantor	Non-woven polyester mat with stamped honeycomb pattern	Polyester / stamped "cells" filled with microspheres
Atvantage CSC and CIC	National Non-wovens	Laid at specified angles and then needled for z-reinforcement	Discontinuous 3–6 inch (76–152 mm) aramid fibers or aramid fiber blends

Note: Lantor Coremat is for open molding only, while Lantor Soric is for closed molding.

Design and Analysis

Sandwich structures are used to provide a combination of light weight, high strength and high stiffness. Two examples of effective sandwich structures include building walls and space mirrors:

- When used as a wall, honeycomb-cored building panels act as a column and will carry loads up to ten times that of conventional frame construction.
- The benefit of using honeycomb core in the construction of large mirrors for space is that the deformation of the honeycomb under gravity is similar to that of a solid blank of the same dimensions, but the honeycomb has only one-fifth the mass of a solid blank.

Cored construction achieves this by acting like an I-beam, with the face skins acting like the flanges and the core functioning as the web.

I-beams are based on the design principle that, in bending, the largest part of the load is carried near the extreme fibers (outer surfaces) of the beam, and very small bending stresses are developed near the neutral axis, as shown

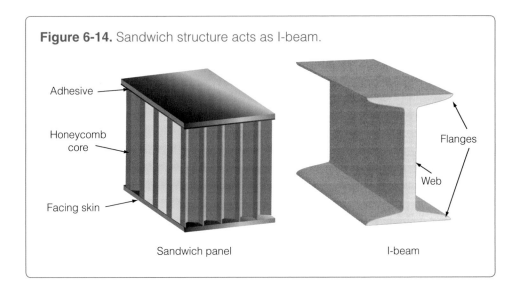

Figure 6-14. Sandwich structure acts as I-beam.

Adhesive

Honeycomb core

Facing skin

Sandwich panel

Flanges

Web

I-beam

in Figure 6-15 below. Notice that the section modulus (stiffness for a given cross-sectional shape) of a rectangular beam is .167 times its height, while an I-beam is more than twice as stiff for the same height. The ideal beam cross-section is where the area is equally split between the top and bottom surfaces, with no material in the middle.

Thus, the top and bottom face skins of a cored sandwich carry most of the load, including in-plane bending, tension and compression. A general rule-of-thumb for lightweight sandwich panels is for face skins to comprise 60–67% of the total panel weight. Many aircraft flooring panels have been optimized for weight savings and have 50% of their weight in the facings and 50% in the core and adhesive.

The core reacts to shear loads and resists out-of-plane compression (e.g., impact) while providing continuous, wide-span support for the face panels. Honeycomb cores have directional properties; thus, careful attention is required to make sure that this is accounted for during design and that the core is placed within the structure correctly to optimize overall structural performance.

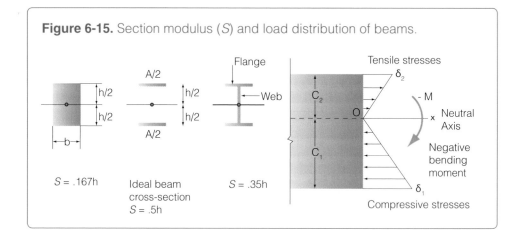

Figure 6-15. Section modulus (S) and load distribution of beams.

General Design Criteria

Some basic criteria to consider when designing sandwich structures include:

- Facings should be thick enough to handle the tensile, compressive and shear stresses caused by the design load.

- Core should be strong enough to withstand the shear stresses caused by the design loads. The adhesive must also have enough strength to carry shear stress into the core.

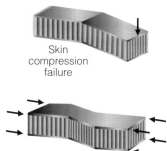

Skin compression failure

- Core should be thick enough and have sufficient shear modulus to prevent overall buckling of the sandwich under load, and also to prevent crimping.

- Compressive modulus of the core and of the facings should be sufficient to prevent wrinkling of the faces under design load.

- Compressive strength of the core must be sufficient to resist crushing by design loads that are normal to the panel facings and also by any compressive stresses induced through flexure.

- Sandwich structure should have enough flexural and shear rigidity to prevent excessive deflections under design load.

Excessive deflection

- The cell size of honeycomb core must be small enough to prevent buckling within the cells.

General Design Guidelines

1. Define loading conditions, including uniform distributed loads, end loading, point loading and possible impact loads.

2. Define panel type (e.g., cantilever, simply supported). Note that fully supported can only be considered when the supporting structure acts to resist deflection under the applied loads.

3. Define physical/space constraints including weight limit, thickness limit, deflection limit and factor of safety.

4. Preliminary calculations:
   - Assume skin material, skin thickness and panel thickness. Ignore core for now.
   - Calculate stiffness and deflection (ignoring shear deflection).
   - Calculate face skin stress and core shear stress.

5. Optimize design. Modify thickness and material selection for both skin and core, if necessary, to achieve desired performance.

6. Detailed calculations:
   - Stiffness.
   - Deflection, including shear deflection.
   - Face skin stress.
   - Core shear stress.
   - Check for the failure modes listed in Design Criteria above, where each is applicable.

Key Property Tests for Sandwich Panels

Long Beam Flex

This is the standard test for determining the load bearing capability of a sandwich panel. It tests the face skins—which should fail before the core—and verifies how much weight the panel will support as well as how much deflection it will experience.

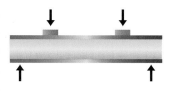

Core Shear

This test reduces the span from the flex test to 15 – 30 times the panel thickness in order to determine where the core will fail before the face skins.

Flatwise Compression

This measures the strength of the core in resisting compressive loads.

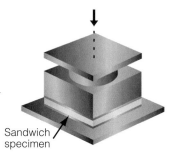

Sandwich specimen

Plate Shear

This measures the shear strength and shear modulus of the core.

Flatwise Tensile

This test measures the strength of the adhesive bond between the face skin and core, and is useful in determining surface preparation for bonding and prepreg adhesion.

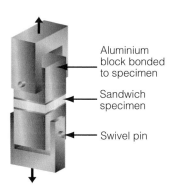

Aluminium block bonded to specimen

Sandwich specimen

Swivel pin

Smoke, Toxic Emissions and Heat Release

Most sandwich panels are used in applications that have some type of smoke, emissions and/or heat release requirements. The test methods and requirements vary greatly depending on whether they are governed by the FAA, or by standards used in the rail industry, bus industry, etc. (*See* Page 156, **Fire, Smoke and Toxicity (FST) Requirements and Heat Release Testing** in Chapter 8.)

If cored structures are being fabricated without using an autoclave, it is usually advisable to vacuum bag the core using an appropriate adhesive or core bonding putty. Core manufacturers can usually provide core bonding guidelines and recommended adhesives.

If using an autoclave, selecting the proper adhesive is vital. For example, a film adhesive developed to provide a good fillet between the face skin and honeycomb cell wall is mandatory when bonding prepreg face skins to honeycomb core.

Cored panels typically need to be sealed or closed off at edges and fastener holes. A variety of edge closure designs are possible, depending on aesthetic and attachment requirements. Core may be routed out at the edges and filled with potting compound or replaced with a more solid material or specially fabricated fitting. Fabricated "z"-sections or "u"-sections may also be bonded to the exterior of the panel edge. Core may also be machined to provide a tapered edge; however, this requires close tolerance work during bonding. Fastener holes typically need to be filled with potting compound or otherwise sealed to prevent moisture ingression.

Figure 6-16. Sandwich panel edge closure designs.

(Courtesy of Hexcel)

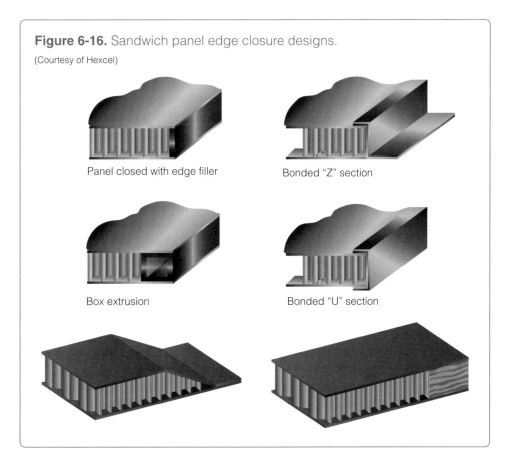

Panel closed with edge filler

Bonded "Z" section

Box extrusion

Bonded "U" section

In-Service Use

Two of the most serious considerations for in-service use of cored structures are impact and moisture absorption. These factors should be considered during the design stage, as careful selection of materials and design can prevent costly maintenance issues and even structural failures.

Aircraft manufacturers have learned to think about such issues as where service vehicles will contact the aircraft (e.g., baggage carts impacting doorways) and the frequency and magnitude of point loading on interior flooring (e.g., doorways and aisles are obvious, but the small strip between seats where shoe heels repeatedly land is equally problematic).

Core has been used successfully in marine structures for over thirty years, so the issue of moisture is not so much environment, as making sure to plan for in-service conditions. Boat builders typically use core bonding putties and vacuum bagging of core materials, but also carefully select the density and type of core depending on whether it is being used in a below waterline, forward section of a hull subjected to slamming from waves, or in a flat deck structure subjected to high heat and a range of compressive and shear loads. Aircraft manufacturers have learned that very thin skins over honeycomb core can result in early failures in structures due to moisture ingress and freeze-thaw cycling, if the proper film adhesives and exterior sealing systems are not utilized.

Note: Honeycomb illustrations in this chapter, including those on Pages 125–127 and 134–137 have been re-created from originals that are copyright Hexcel Corporation, and the information in them is reproduced courtesy of Hexcel (all rights reserved).

Introduction to Tooling

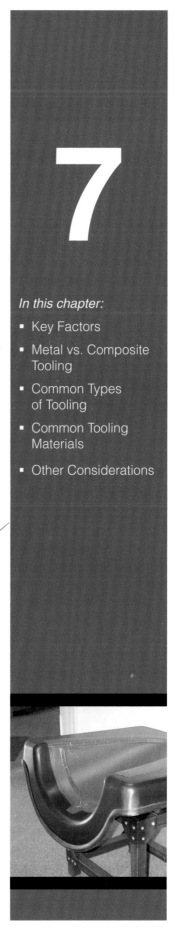

7

Key Factors

Advanced composite materials are unique in the sense that they are molded to a desired shape. Unlike metals that require bending, forming, casting, and/or machining into shape, composites are molded. A layup mold (or bond fixture) is designated for this purpose. These "tools" must produce a dimensionally accurate part, often through an elevated temperature process.

In addition to the primary molding processes, tooling may be required to facilitate many post-molding operations such as trimming, drilling, bonding, fastening, and assembly of composite structures. Good quality precision tooling is the key to reliable and repeatable production. Without tooling, every part is a one-of-a-kind piece.

Production Requirements

Tools are designed to meet the production quantities that are outlined for the project. A tool designed to make twenty parts may be very different from one designed to make 20,000 parts. The production *rate* requirements often drive the choice of which processes will be used to make the parts and ultimately the tool design and the tooling costs.

Tool Function and Types of Tools

For any given composite part or assembly, a series of tools are required to control certain aspects of the manufacturing process. Each tool is designed around how the tool will function. For example, a mold that might be used for hand layup of prepreg materials would be required to control only one surface of the

In this chapter:
- Key Factors
- Metal vs. Composite Tooling
- Common Types of Tooling
- Common Tooling Materials
- Other Considerations

139

part and accommodate a vacuum bag so as to allow the operator to layup, bag, and cure a dimensionally accurate part.

Compared to a mold used for resin transfer molding (RTM), where the tool must control both surfaces of the part, it must allow for the positioning and clamping of a preform, have ports for resin injection, and a vacuum-assist cavity or a vent. The tool function is often described in the tool description or name. Many companies use an abbreviated description or a tool code to designate each tool and its function.

Some common types of tools used in conventional fiber-reinforced thermoset matrix composite manufacturing:

- Layup Mold or Mandrel (LM)
- Trim Fixture or Jig (HRF, HRJ, or TF)
- Drill Fixture or Jig (DF or DJ)
- Bond Assembly Fixture or Jig (BAF or BAJ)
- Combination Fixture (CF or CJ)
- Semi-rigid Tools and Cauls (Caul)
- Elastomeric Vacuum Tool (EVT)

Tooling Material Properties

The material used to construct tooling is critical to the performance of the tool, especially when the tool is subjected to elevated temperatures. For example, aluminum has a high coefficient of thermal expansion (CTE) compared to that of a carbon fiber-reinforced epoxy laminate. Because of this, care must be taken when selecting aluminum for a mold or mandrel that will be used to make parts. Aluminum does have very good heat transfer properties that may be advantageous in an RTM or press-molding process. A thorough understanding of material properties is required to design functional tooling. Typical materials used in tool construction include:

Composite Materials

- Glass fiber reinforced plastic (GFRP)
- Carbon fiber reinforced plastic (CFRP)
- Carbon fiber reinforced elastomer (CFRE)

Metals

- Electro-formed nickel alloy (EFNi)
- Aluminum
- Steel (stainless and other tool steel alloys)
- Invar 36 or 42

TABLE 7.1 General Comparison of Tooling Material Thermal Properties

Material	Specific Gravity	Specific Heat (Btu/lb/°F)	Thermal Mass (Btu/lb/°F)	Thermal Conductivity Coefficient (Btu/ft²/hr/°F/in)	Coefficient of Thermal Expansion (in/in/°F)
Aluminum	2.70	0.23	0.62	1395	12.9×10^{-6}
Stainless Steel	8.02	0.12	0.96	113	9.6×0^{-6}
EF-Nickel	8.90	0.10	0.89	500	7.4×10^{-6}
Invar 36	8.11	0.12	0.97	72.6	0.8×10^{-6}
GFRP	1.80-1.90	0.3	0.54-0.60	21.8-30.0	$8.0\text{-}9.0 \times 10^{-6}$
CFRP	1.50-1.60	0.3	0.45-0.48	24.0-42.0	$0.0\text{-}6.0 \times 10^{-6}$

Calculating the Coefficient of Thermal Expansion (CTE)

Example:

Carbon fiber tool laminate CTE (in/in/°F) = 2×10^{-6} = 0.000002

Aluminum tool CTE (in/in/°F) = 12.9×10^{-6} = 0.0000129

Cure temperature = 350°F

Desired part length = 48.00 in

Calculating the Effect:

Temperature differential (cure temp - ambient):

DT = 350°F - 70°F = 280°F

DT = 177°C - 21°C = 138°C

Relative Expansion:

Carbon fiber tool, 0.000002 in/in/°F x 280°F = 0.00056 in/in

Aluminum tool, 0.0000129 in/in/°F x 280°F = 0.0036 in/in

Total Expansion:

Carbon fiber tool, 0.00056 in/in x 48.00 in = 0.027 in

Aluminum tool, 0.0036 in/in x 48.00 in = 0.173 in

After-effect of temperature on a carbon fiber reinforced plastic (CRFP) tool/part

After-effect of temperature on an aluminum tool/CRFP part

Thermal Conductivity

Thermal conductivity is the ability of a material to conduct heat. Aluminum and copper are very good thermal conductors. Ceramic is not a good thermal conductor. Carbon fiber reinforced plastic (CFRP) has a higher thermal conductivity coefficient (T.C.C.) than glass fiber reinforced plastic (GFRP) tooling. (Glass tends to be a good insulator.) Thus CFRP tooling materials will conduct heat more efficiently than GFRP Materials.

Thermal mass is the amount of energy required to heat a given material is related to the density and the specific heat of the material. This is known as the thermal mass. The thermal mass of the tooling affects its ability to heat and/or cool at the rates required by the process specifications. As a result, tools made with heavy materials (metals) tend to heat more slowly than tools made of CFRP or GFRP materials.

Metal vs. Composite Tooling

Metals

Tools designed for high-temperature cure cycles in large production quantities are usually made from metal. In aerospace, Invar 36 or 42 alloys are used for this purpose. These tools are very durable, and can be machined accurately to tight tolerances. However, they tend to be very heavy, and the high thermal mass and low TCC can lead to very slow heat-up rates for the prepreg plies next to the tool face (as opposed to those next to the bag side). This can lead to problems in processing.

Metal (with the exception of Invar) tools can also have very different expansion rates than composites, leading to problems with dimensional accuracy in the cured parts, since they cure at the expanded dimensions of the hot tools. This may lead to physical stress on the part and difficulty removing the part from the mold. Conversely, this characteristic may help when molding parts such as golf clubs, as the internal metal mandrel will contract on cool down, pulling away from the part.

Composites

Composite tools solve many of the metal tool problems, while introducing different problems of their own. They are much lighter, and the low thermal mass helps achieve desired ramp rates. The lighter weight makes them much

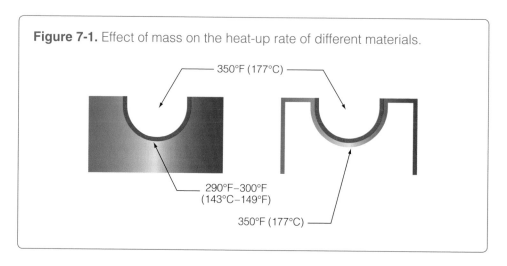

Figure 7-1. Effect of mass on the heat-up rate of different materials.

350°F (177°C)

290°F–300°F (143°C–149°F)

350°F (177°C)

easier to move around in the shop, and may in some cases eliminate the need for a forklift or crane to move small to medium-sized tools.

Composite tools are fairly easy to duplicate. Multiple units of the same, exact mold can be made from a single pattern. This is an advantage over metal tooling, as the duplicates are dimensionally identical to each other and are not affected by machine setup errors or compounding setup tolerances.

On the down-side, composite tools are easier to damage than metal tools, and will ultimately develop microscopic cracks (microcracks) in the tool laminate due to differing rates of thermal expansion between the fibers and the matrix. These cracks in a layup mold will eventually grow together through enough thermal cycles, and cause vacuum leaks throughout the tool. This can be detrimental to the process of vacuum-bagged or autoclaved prepreg composite parts.

For this reason, composite tools have a limited production life compared to metal tools. It is very unusual to get over 200–250 cycles from a composite tool before the leak rate becomes noticeable or even unacceptable. Often the number of cycles is much less, down to just a few cycles in particularly bad cases, depending on the tool design, quality of the materials, and workmanship used in the construction of the mold laminate. Composite tools can be made with an internal fluoroelastomeric membrane that can eliminate the

Figure 7-2. Composite tooling for CFRP "Turtle Back" fairing—layup mold (LM) along with CFRP sandwich panel after processing.

Layup mold, trim and drill fixture (TDF), core forming jig (CFJ).

Potted and formed honeycomb core assembly, and a core kit template (CKT) used to layout and pot the core.

leak potential in the tool laminate, but this technology is not well known and highly under-used within the aerospace industry.

Therefore composite tools are not normally used in "high-rate" production, such as in the automotive business, as production quantities are just too high to make composite tooling practical. However, in the aerospace or marine industries, where production quantities run from tens to hundreds, composite tools are quite often the best choice.

With proper planning, multiple units of composite tools may prove to be less expensive and more productive than multiple units of metal tools. Molds made from composite materials tend to be more robust when designed with self-supporting (monocoque) features that also move the damage-prone edges away from the part layup/vacuum bag seal area. (*See* Figure 7-3.)

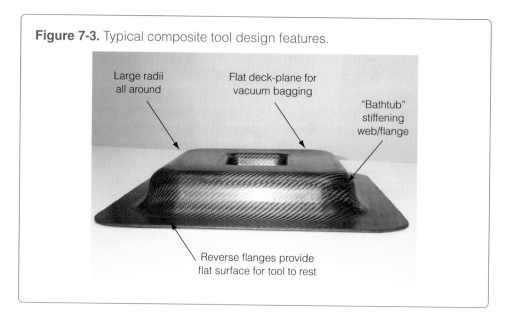

Figure 7-3. Typical composite tool design features.

Large radii all around

Flat deck-plane for vacuum bagging

"Bathtub" stiffening web/flange

Reverse flanges provide flat surface for tool to rest

Tooling for Thermoplastics

Tooling Materials

- Steel
- Nickel (EFNi)
- Invar 36 or 42
- CFRP—made with BMI or Cyanate Ester resins
- Monolithic graphite
- Wash-away ceramics
- Cast ceramics
- Reinforced ceramics

Other Considerations

- Weight, transportation, and handling
- Operator friendly: simple, uncomplicated design/fabrication/usage
- Protection from Operator Induced Damage (O.I.D.)
- Protection from Forklift Induced Damage (F.L.I.D.)
- Cost, both initial tooling cost and the amortized cost per part in production
- Designing part removal features and methods in the molds

Final Note on Tool and Part Design

As composite engineers begin to design larger and more complex monocoque structures, the need for more innovative tooling and manufacturing methods will emerge. Building multi-piece molds for co-curing entire assemblies will require traditional "tool designers" to become "mold-designers," thinking in terms of eliminating as much of the secondary assembly as possible.

This shift in approach requires that the industry also recognize that extra costs associated with designing, building, and using these complex molds are recognized as being cost effective in the long run. These molding costs can easily be offset by the fact that there will be far less assembly and assembly tooling required to make larger molded structures. As a result, the cost of inspection and quality assurance is also shifted in this direction toward the **layup-assembly** approach and away from the layup, machine, and then assemble approach.

8

Inspection and Test Methods

Destructive Coupon Testing

Any claims as to the properties of a given composite material must be validated through destructive coupon testing that (1) is performed according to an accepted standardized method and (2) is supported by statistical data analysis. Tests developed for metals and plastics typically cannot be used for composites. Standard methods specifically for testing composites have been developed by the American Society of Testing and Materials (ASTM), the International Standards Organization (ISO), and a variety of composites and plastics industry associations as well as materials manufacturers. Table 8.1 below shows a typical composite material test matrix, in which all of the tests are performed according to specific ASTM test methods, and statistical analysis is provided by testing 6 coupons for each property and environmental condition.

In this chapter:

- Destructive Coupon Testing

- Resin, Fiber and/ or Void Content

- Fire, Smoke and Toxicity (FST) Requirements and Heat Release Testing

- Non-Destructive Testing

TABLE 8.1 Typical Composite Material Test Matrix

Mechanical Property	Test Method	Test Condition and Number of Tests per Batch			No. of Tests
		Min.Temp. Dry	Room Temp. Dry	Max.Temp. Wet	
0° tension	ASTM D3039	6	6	6	90
90° tension	ASTM D5450	6	6	6	90
0° compression	ASTM D3410	6	6	6	90
90° compression	ASTM D5449	6	6	6	90
In-plane shear	ASTM D5448	6	6	6	90
Interlaminar shear	ASTM D5379	6	6	6	90
				TOTAL	**540**

Courtesy of ASM Handbook Volume 21 Composites.

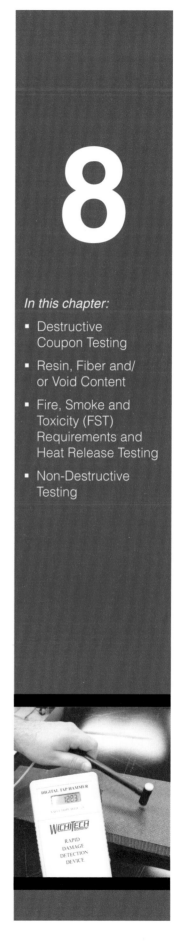

147

The goal of the overall test program is to generate data that will represent the actual composite structure being built, throughout its operational life, taking into consideration all possible environmental conditions. Much work has been done on developing accurate test methods for composites, but much debate still remains. Issues such as specimen preparation, environmental conditioning, test setup and instrumentation are all vital to obtaining accurate and repeatable data. The discussion here will present a general overview of tests typically performed to generate mechanical property data for composites, which includes the following lamina (single-ply) and laminate (two or more plies) testing categories:

- Tensile
- Compressive
- Shear
- Flexure
- Fracture toughness
- Fatigue

Table 8.2 shows the mechanical properties and test methods from which MIL-Handbook-17 accepts data on composites.

TABLE 8.2 Standard Test Methods from which MIL-Handbook-17 Accepts Data

Lamina/Laminate Mechanical Test	Test Methods
0° tension	ASTM D3039
90° tension	ASTM D3039, D5450
0° compression	ASTM D3410, D5467, D6641
90° compression	ASTM D3518, D5449, D6641
In-plane shear	ASTM D3518, D5448, D5379
Interlaminar shear	ASTM D5379
Short beam strength	ASTM D2344
Flexure	ASTM D790*
Open-hole compression	ASTM D6484
Open-hole tension	ASTM D5766
Single-shear bearing	ASTM D5961
Double-shear bearing	ASTM D5961
Compression after impact	ASTM D7137
Mode I fracture toughness	ASTM D5528
Mode II fracture toughness	ASTM D6671
Tension/tension fatigue	ASTM D3479

* Not included in the original table but described on Pages 151-152.

Courtesy of ASM Handbook Volume 21 Composites.

Tensile

Tensile tests are basically achieved using a rectangular coupon that is held at the ends by wedge or hydraulic grips and then loaded in uniaxial tension using a universal testing machine. The coupon may be straight-sided for its entire length or it may be width-tapered from the ends into the gage section (often called a *dog-bone* or *bow-tie* coupon). The gage section, or gage length, is the part of the specimen over which strain or deformation is measured. In tensile testing, this material response is measured using strain gages or extensometers. Changing the coupon configuration enables composites with woven and multiaxial reinforcements to be evaluated in addition to the unidirectional composites typically tested using straight-sided coupons. End tabs are sometimes bonded onto the specimen in order to better distribute the load and prevent stress concentrations. Test results include tensile modulus (usually measured in Msi or GPa) and tensile strength (usually measured in Ksi or MPa).

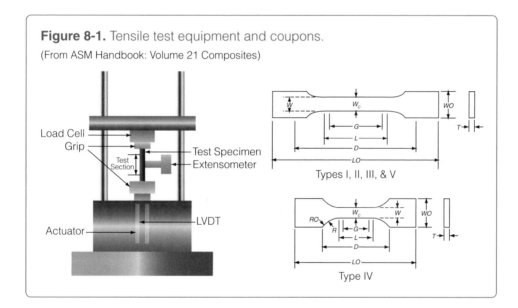

Figure 8-1. Tensile test equipment and coupons.
(From ASM Handbook: Volume 21 Composites)

Compressive

Compression tests use a flat, rectangular coupon, held by tapered wedge grips that may be conical, rectangular or pyramid-shaped, and may introduce load into the specimen through shear, end-loading, or a combination of shear and end-loading. Compression tests may also use a supported test section, in which the specimen faces are supported by a test fixture, or an unsupported test section, in which the specimen faces are unrestrained. After placing the coupon in the text fixture and the whole assembly into a universal testing machine, compressive load is transferred to the coupon through the grips and the material response is measured. Results include compressive modulus and compressive strength.

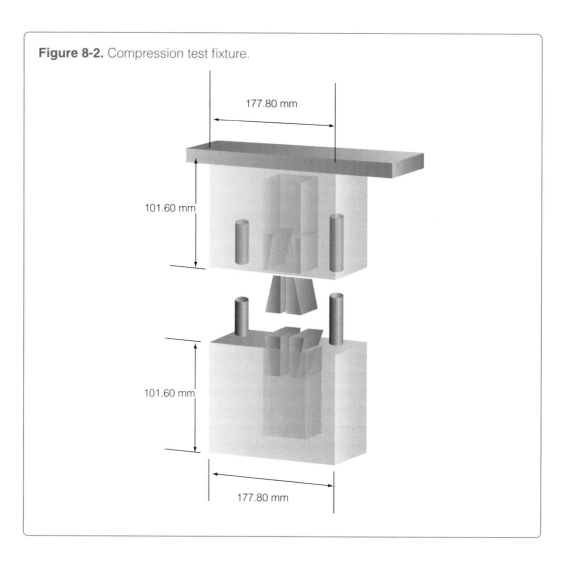

Figure 8-2. Compression test fixture.

177.80 mm

101.60 mm

101.60 mm

177.80 mm

Shear

Shear testing of composite materials has traditionally been a very difficult area in which to define a consistently accurate test. Shear modulus has been much less problematic than shear strength, where issues such as edge effects, nonlinear behavior of the matrix or the fiber-matrix interface, imperfect stress distributions, and normal stresses have made resulting data questionable.

ASTM D3518 uses a modified ASTM D3039 tensile test coupon with a fully balanced and symmetric material layup of ±45°. The test is conducted similarly to D3039 using a tensile test fixture and testing machine and the resulting strain measurements give in-plane shear modulus and in-plane shear strength.

ASTM D5379 is often referred to as the V-Notched Beam Method or Iosipescu Shear Test (after one of its founders) and uses a flat rectangular coupon with symmetrical V-notches on opposing sides of the coupon at is midpoint (*see* Figure 8-3). The coupon is loaded through a specially designed fixture where the upper head of the fixture is attached to and pushed downward by the cross-head of the testing machine. The fixture thus introduces a shear load into the coupon between the two notches, and deformation is measured by

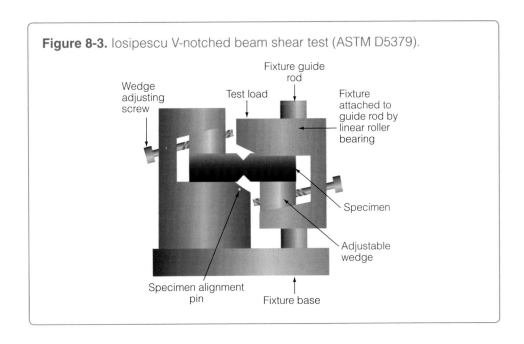

Figure 8-3. Iosipescu V-notched beam shear test (ASTM D5379).

strain gages. Tabs may be bonded to the ends of the coupon for increased stability, and in-plane or out-of-plane shear may be measured, depending on the orientation of the material relative to the loading axis.

Out-of-plane shear may also be measured by ASTM D2344, commonly known as the Short-Beam Shear (SBS) Test. This test attempts to measure the interlaminar (out-of-plane) shear strength of parallel-fiber-reinforced composites and uses a short, deep beam cut from a flat laminate panel. The short, deep beam is meant to minimize bending stresses while maximizing out-of-plane shear stresses. The coupon is mounted as a simply supported beam and loaded at the midpoint. However, contact stresses introduced at the load points interfere with the strain distribution, giving a failure that is rarely pure shear in nature. Thus, it has been recommended that this test be used only for qualitative purposes.

Flexure

ASTM D790 was originally developed for unreinforced plastics but has been modified and approved for composites. It measures the force required to bend a beam, and measures flexural strength, flexural stress at specified strain levels, and flexural modulus can be calculated as follows:

- **Flexural modulus:** The ratio of outer fiber stress to outer fiber strain.
- **Flexural stress at yield:** The outer fiber stress corresponding to test specimen yield.
- **Flexural stress at break:** The outer fiber stress corresponding to test specimen failure.
- **Flexural strength:** The maximum outer fiber stress sustained by a specimen during testing.

The rectangular beam specimen is fixture as a simply supported beam and subjected to three-point bending by applying load to its midpoint. The most

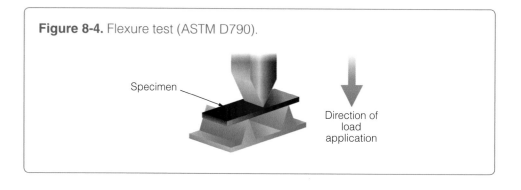

Figure 8-4. Flexure test (ASTM D790).

Specimen

Direction of load application

commonly used specimen size is 3.2mm x 12.7mm x 64mm (0.125" x 0.5" x 2.5") for ASTM and load is applied at a specified rate as follows: (1) 0.01 mm/mm/min for materials that break at small deflections; and (2) 0.10 mm/mm/min for materials that undergo large deflections during testing. The specimen is deflected until it either breaks or the outer fiber strain reaches 5%. However, flexural strength cannot be determined if the coupon does not break or if it does not fail in its outer surface within the 5.0% strain limit. ASTM D6272 is an alternative method using a four-point loading system.

Since the physical properties of materials (especially thermoplastics) can vary depending on temperature, a thermal chamber may be used to test materials at temperatures that simulate the in-service environment. Standard test fixtures are installed inside the chamber, and testing is conducted the same as it would be at ambient temperature. The chamber has electric heaters for elevated temperatures and uses carbon dioxide or other gas as a coolant for reduced temperatures.

Fracture Toughness

Fracture in a composite structure is usually initiated by a crack or flaw which creates a local stress concentration. Fracture toughness is a generic term for the measure of resistance to extension of a crack or initial flaw. As shown in Figure 8-5, fracture toughness testing can be one of three types, depending on the mode in which the crack propagates:

- Mode I, Opening/tensile mode
- Mode II, Sliding/shear mode
- Mode III, Tearing mode

ASTM D5528, "Standard Test Method for Mode I Interlaminar Fracture Toughness of Unidirectional Fiber-Reinforced Polymer Matrix Composites" is the most common standardized fracture toughness test for composites, and is often called the Double-Cantilever Beam Test. This test method is limited to unidirectional carbon fiber and glass fiber tape laminates because use of woven and multiaxial reinforced composites may increase the tendency for the delamination to grow out-of-plane, compromising the accuracy of the test results. The coupon is sized 125 mm long, 20 to 25 mm wide, and 3 to 5 mm

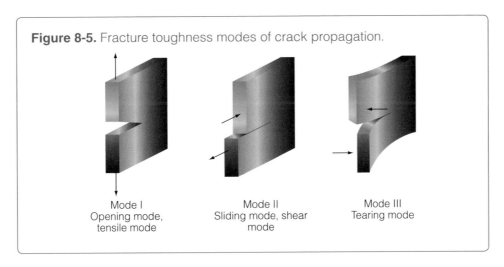

Figure 8-5. Fracture toughness modes of crack propagation.

Mode I
Opening mode,
tensile mode

Mode II
Sliding mode, shear
mode

Mode III
Tearing mode

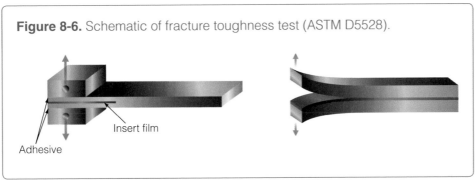

Figure 8-6. Schematic of fracture toughness test (ASTM D5528).

Insert film

Adhesive

thick (5 inches long, 0.8 to 1 inch wide, and 0.12 to 0.20 inches thick). A non-bondable film insert is placed at the mid-plane during manufacture in order to create an initial crack length of roughly 63 mm (2.5 inches).

Fatigue

Most structures do not operate under a constant load. In fact, often stresses on a composite structure are constantly changing. A good example is a rotating shaft such as the axle on a railroad car. The bending stresses change from tension to compression as the axle rotates. This constant change in stress can cause fatigue failure in which the material suddenly fractures. The process behind this is the initiation and growth of cracks in the material. Other examples of cyclic loading include the forces on an airplane wing. Air turbulence, changing wind directions and changing speeds, for example, can combine to make it nearly impossible to determines the stresses acting on an aircraft wing at any given instant in time. Thus, a complete fatigue testing program for such a structure is comprised of statistical analysis where the worst-case states of stress (for example: landing, take-off and turbulence), as well as the mean and amplitude stresses, are measured and factored in order to determine the long-time fatigue behavior.

Figure 8-7. S-N Curves for [02/902]S Cross-ply Laminates HTA/F922 carbon/epoxy *(left)* and E-glass/ F922 epoxy *(right)*.

Carbon/epoxy cross-ply laminates show excellent fatigue resistance. The fatigue life Nf exceeded 4 x 106 cycles at loads approaching 90% and 70% of the ultimate tensile strength (UTS) for [0/90]4S T300/924 and 70% UTS for [02/902]S, respectively. End tabs are prone to debond during testing; thus, prevention of end tab debonding is essential for ensuring reliable fatigue data.

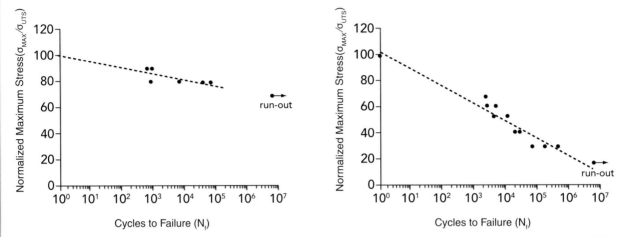

From *NPL Report Measurement Note CMMT(MN) 067*, "Environmental and Fatigue Testing of Fibre-Reinforced Polymer Composites," December 2000, by W. R. Broughton and M. J. Lodeiro, Industry and Innovation Division, National Physical Laboratory, Middlesex, U.K.; used with permission.

To measure the fatigue resistance of materials, a cyclic load is applied to a specimen until it breaks. The number of cycles to failure, called the fatigue life (N_f) is recorded. The logarithm of this life is plotted against the stress (or sometimes the log of stress) to develop an S-N curve (stress versus number of cycles to failure) as shown in Figure 8-7. This curve defines the fatigue resistance of a material. As can be seen in the figure, there is a great deal of scatter in the fatigue life. Consequently, a large number of tests must be run in order to define the fatigue resistance.

ASTM D3479, "Standard Test Method for Tension-Tension Fatigue of Polymer Matrix Composite Materials," is used to determine the fatigue behavior of polymer matrix composite materials subjected to tensile cyclic loading. This test method is limited to un-notched test specimens subjected to constant amplitude uniaxial in-plane loading. Two different procedures are used:

- **Procedure A:** The servo-hydraulic test machine is controlled so that the test specimen is subjected to repetitive constant amplitude load cycles; i.e., constant load.
- **Procedure B:** The test machine is controlled so that the test specimen is subjected to repetitive constant amplitude strain cycles; i.e., constant strain.

Other test parameters are listed in Table 8.3.

TABLE 8.3 Standard Test Methods from which MIL-Handbook-17 Accepts Data

Fatigue Test Parameter	Description
Upper cycle count limit or "Runout"	It is possible to set the maximum load low enough to cycle a specimen indefinitely before failure. Thus, a runout is typically set at which the residual strength of the specimen may be measure statically if the specimen does not fail.
R-ratio	Ratio of the minimum to maximum applied stress or strain. A tension-tension fatigue test with stresses cycling from 10 to 100 kN (2,250 to 22,500 lb) has an R-ratio of 0.1. A reverse-loaded test (tension-compression) with stresses of 50kN to -50kN (11,240 to -11,240 lb) has an R-ratio of -1.0.
Frequency	Best practice is to use the same frequency for each specimen, ensuring that the frequency used is not so fast that the specimen heats excessively (generally accepted as greater than 20°F/10°C above ambient).
Waveform	Most fatigue tests apply load using a sinusoidal waveform.

Courtesy of ASM Handbook Volume 21 Composites.

Resin, Fiber and/or Void Content

There are two basic methods for determining the resin, fiber and/or void content in a composite:

- Matrix Ignition Loss (Burn-out)
- Matrix Digestion (Acid Digestion)

Matrix Ignition Loss

The two most common test methods used are ASTM D2584 and ASTM D3171. ASTM D2584, "Standard Test Method for Ignition Loss of Cured Reinforced Resins," uses a pre-dried and weighted composite specimen, which is placed in a crucible of known weight and then inserted into a furnace, where the organic resin matrix will burn off. After burning, the crucible is cooled in a dessicator and reweighed. Ignition loss is the difference in weight before and after burning, and represents the resin content of the composite sample. The remaining weight determines the fiber reinforcement and any inorganic fillers in the composite. The fiber volume fraction is calculated as:

$$V_f = \frac{\rho_m W_f}{\rho_f W_m + \rho_m W_f}$$

where:

V_f = volume fraction of fibers

W_f = weight of fibers

W_m = weight of matrix

ρ_f = density of fibers

ρ_m = density of matrix

This method assumes that the fiber reinforcement is unaffected by combustion and is commonly used with fiberglass-reinforced composites, as well as other composite reinforcements that are insensitive to high temperatures such as quartz (silica).

Matrix Digestion

ASTM D3171 is commonly referred to as the *acid digestion* method, but it is actually described by ASTM as "Standard Test Methods for Constituent Content of Composite Materials." Designed for two-part composite material systems, this standard uses one of two approaches:

- **Method I**. The matrix is removed by digestion or ignition using one of seven procedures, leaving the reinforcement essentially unaffected and enabling calculation of the reinforcement or matrix content (by weight or volume) as well as percent void volume. Procedures A through F are based on chemical removal of the matrix while Procedure G removes the matrix by igniting the matrix in a furnace.

- **Method II**. Applicable only to laminate materials of known fiber areal weight, and calculates the reinforcement or matrix content (by weight or volume), as well as the cured ply thickness, based on the measured thickness of the laminate. Void volume is not measured using this method.

To determine the fiber volume fraction using acid digestion, the matrix is dissolved using an appropriate solvent and the remaining fibers are then weighed. Alternatively, a photomicrographic technique may be used in which the number of fibers in a given area of a polished cross section is counted and the volume fraction determined as the area fraction of each constituent. Chemical digestion assumes a chemical has been selected that does not attack the fibers. Common choices are hot nitric acid for carbon fiber/epoxy, sulfuric acid and hydrogen peroxide for carbon fiber/polyimides and PEEK, and sodium hydroxide for aramid composites. Others are described in the ASTM D3171 specification.

The void volume of the composite is calculated, where possible, by using the known composite density, cured resin density, fiber density, and determined resin and fiber contents by weight. The density properties are usually available from the materials supplier and should be obtained for the specific lot of material being tested. ASTM D2734 "Void Volume of Reinforced Plastics" specifies these calculations in detail.

Fire, Smoke and Toxicity (FST) Requirements and Heat Release Testing

Note: Text from this section adapted from "Passenger Safety: Flame, Smoke Toxicity Control" by Jared Nelson, *Composites Technology Magazine*, December 2005: reproduced by permission of *Composites Technology Magazine*, copyright 2005, Gardner Publications, Inc., Cincinnati, Ohio, USA.

Composite structures must be tested to meet fire, smoke, toxicity (FST) and heat release requirements, according to their application. The regulations and standards for composites used in commercial aircraft differ greatly from those used in rail and bus, ships, submarines, military aircraft, and the construction/building industry. Fireworthiness and fire safety standards also vary greatly by country.

Fire properties of composite materials measure their ability to ignite, burn and generate toxic substances. The properties commonly tested include ignition time, heat release rate, heat of combustion, flame spread, smoke density and smoke toxicity. The FST and heat release tests used for commercial aircraft and rail applications in the U.S. are discussed in this section.

Commercial Aircraft

In 1988, the Aviation Safety Research Act was passed, mandating that the Federal Aviation Administration (FAA) conduct long-term investigations into aircraft fire safety, including fire containment and the fire resistance of materials in aircraft interiors. One of the goals of this program is to enable the design of a totally fire-resistant cabin for future commercial aircraft. Thus, FAA regulations are continually changing with technology, trending toward more stringent standards and tests more reflective of actual service. For example, the FAA regulation for non-metallic materials in pressurized aircraft cabins have been reduced from a total and peak heat release of 100 kW/m^2 to 65 kW/m^2. An upper limit on smoke density has also been enforced.

Current FAA FST requirements are described in Title 14 of the Code of Federal Regulations (14 CFR) Part 25, specifically in regulation §25.853 and Appendix F.

14 CFR 25.853a is a vertical Bunsen burner test—commonly referred to as vertical burn—to determine the resistance of cabin and cargo compartment materials to flame. A specimen is held in a vertical position by a device inside a cabinet and a Bunsen burner is placed beneath it for 60 seconds (i) or 12 seconds (ii). The burner is then removed and the specimen is observed. The following data are recorded as shown in Tables 8.4 and 8.5.

TABLE 8.4 Data Recorded for 14 CFR 25.853 Vertical Bunsen Burner Test

Ignition Time	Length of time burner flame is applied to specimen
Flame Time	Time that specimen continues to flame after burner is removed
Drip Extinguishing Time	Time that any flaming materials continues to flame after falling from specimen
Burn Length	Distance from original specimen's edge to farthest evidence of damage to specimen

TABLE 8.5 Requirements for Passing 14 CFR 25.853 Vertical Burn Test

Test	Flame Time (sec)	Average Drip Extinguishing Time (sec)	Average Burn Length
(i) 60 seconds	<15	<3	6 in. (152.4mm)
(ii) 12 seconds	<15	<5	8 in. (203.2mm)

14 CFR 25.853d is comprised of two separate tests: (1) OSU (Ohio State University) Rate of Heat Release; and (2) Specific Optical Density of Smoke Generated by Solid Materials. The OSU Rate of Heat Release is designed to test whether or not materials will flashover (the temperature at which the heat in an area is enough to ignite all flammable material simultaneously), or readily contribute to a fire in a crash situation. It provides information on fire size and growth rate. The rate of heat release measures the rate at which a burning material releases heat, using the principle of oxygen consumption (calorimetry). This small-scale component test, published under ASTM E906, uses the OSU calorimeter and measures a specimen's heat release when exposed to a constant, external radiant heat source producing a heat flux (heat transfer per unit of cross-sectional area) of 35 kW/m^2. Passing criteria can be seen below in Table 8.6.

The smoke density test, previously known as the ASTM F814 (now retired), closely follows ASTM E662, which is often referred to as the "NBS smoke density chamber" (NBS: National Bureau of Standards). This test measures the smoke generated by a burning material expressed in terms of specific optical density (Ds); it is designed to help improve the ability for passengers to exit an aircraft cabin during a fire. Ds readings are taken at 1.5 minutes into the test and 4 minutes. Passing criteria is given in Table 8.6.

TABLE 8.6 Requirements for Passing 14 CFR 25.853d OSU Heat Release Test and Smoke Density Test

Total Heat Release within first 2 minutes	≤ 65 kW·min/m^2
Peak Heat Release	≤ 65 kW/m^2
4-minute Smoke Density (Ds)	≤ 200

ASTM E1678 outlines testing for toxic gas emission and measures the lethal toxic potency of smoke produced from a material which is ignited while exposed to a radiant heat flux of 50 kW/m^2 for 15 minutes. Test specimens are no larger than 3 by 5 inches (76 by 127 mm), with a thickness no greater than 2 inches (51 mm), and are intended to represent actual finished materials or products used, including composite and combination systems. Concentrations of the major toxic gases are measured over a 30-minute period. These measurements are then used in predictive equations to determine the analytical toxic potency of a material, based on data for carbon monoxide, carbon dioxide, oxygen, hydrogen cyanide, hydrogen chloride, and hydrogen bromide. Thus, the test's predictive ability is limited to those materials whose smoke toxicity can be attributed to one or more of these toxicants.

In the U.S., the Federal Railroad Administration of the Department of Transportation (DOT) regulates the safety of passenger trains, buses and other "people movers." Fire safety requirements are found in Title 49 of the Code of Federal Regulations, Chapter II, Part 238.103, wherein Appendix B dictates flame spread and smoke requirements in areas where composites are used based on two standards developed by ASTM.

ASTM E162 "Standard Test Method for Surface Flammability of Materials Using a Radiant Heat Energy Source" measures flame spread. This test is performed using a 15.2 cm by 45.6 cm (6 inch by 18 inch) panel tilted at a 30° angle. The panel's top edge is exposed to a 670°C/1,238°F heat source placed at a distance of 11.9 cm/4.7 inches. The test is run either until the flame reaches the bottom edge, or (if it does not) for 15 minutes. The flame spread index of the material (Is) is calculated based on the distance the flame traveled and the amount of heat generated, with a passing criteria of ≤35.

Smoke density is measured using the same standard as for commercial aircraft, ASTM E662. Smoke toxicity is not currently regulated in 29 CFR II, Part 238.103, but can be measured using either the Boeing Specification Support Standard or the Bombardier SMP 800-C test. Both tests use a smoke chamber. The Boeing test method employs a flamed heat source, to gauge toxic fume concentration. The Bombardier method, however, uses colorimetric tubes or absorptive sampling. In the first instance, colorimetric tubes are placed into the smoke chamber. Each tube contains a specific fluid that reacts with a specific gas, resulting in a change of fluid color. Gas concentration is determined using a color band scale. Alternatively, absorptive sampling uses light spectroscopy. Each gas has a unique spectroscopic signature that permits technicians to identify the gas and determine its level of concentration.

Both tests gauge the concentrations of six gases—carbon monoxide (CO), hydrogen fluoride (HF), hydrogen chloride (HCl), hydrogen cyanide (HCN), nitrogen oxides (NOx) and sulfur dioxide (SO_2). The Bombardier test also tests for carbon dioxide (CO_2) and hydrogen bromide (HBr). As in the smoke density test, concentration levels are recorded at specified time intervals. For the Boeing test, the maximum allowable levels for the six gases are 3,500, 200, 500, 100, 100 and 100 parts per million (ppm), respectively. For the Bombardier test, carbon dioxide and hydrogen bromide are specified at 90,000 and 100 ppm, respectively. Otherwise, permitted maximums are identical to the Boeing test, with the exception of hydrogen flouride, which is reduced to 100 ppm. Ultimately, end-users and local municipalities must determine acceptable toxicity levels.

ASTM E162 and ASTM E662 criteria were originally developed using unsaturated polyester resin as a baseline. Since then, some local municipalities have developed more stringent criteria. After a 1979 passenger railcar fire in San Francisco's Bay Area Rapid Transit (BART) system, it was determined that many interior components (of composite and other materials)—the seats, in particular—did not meet applicable ASTM requirements. Since that time, BART has replaced seating with nonflammable, nontoxic materials, and its cars not only meet but beat ASTM criteria. Its suppliers now must conform to standards that exceed federal requirements. For example, while ASTM E662 specifies the four-minute smoke density at less than 200, BART dictates that the level must remain at 100 or below.

Outside of the U.S., most countries have developed FST requirements similar to U.S. limits. However, following a fire in the Kings Cross station of the London Underground in 1987, the British government implemented much more stringent flame spread and smoke density requirements, which are outlined in British Standard (BS) 6853 and BS 476, respectively. Unlike the U.S. requirements, British requirements specify different criteria depending on the transportation environment. For instance, trains that travel underground have more stringent smoke density requirements than trains that travel above ground.

Since the establishment of ASTM standards, fire-retardant-filled resin systems have been developed that are far more flame/smoke resistant than the polyesters used to establish ASTM flame/smoke criteria. A common solution is the addition of fire retarding materials to neat resins to improve FST performance. Intumescent coatings may also be used (intumescence describes the phenomenon of swelling). Intumescent systems with flame-retarding capacity usually contain three components: an acid, carbon (not carbon fiber), and a catalyst, which interact to form a char layer. This then effectively eliminates an at-risk coated material (e.g., a balsa-cored sandwich panel) as a fuel source for the flame. Recently, intumescent materials have been incorporated into the composites themselves, to take advantage of a char barrier without resorting to secondary coating operations.

Non-Destructive Testing

There are a variety of **non-destructive inspection** techniques available to help determine the extent and degree of damage. Each has its own strengths and weaknesses, and more than one method may be needed to produce the exact damage assessment required.

Visual Inspection

Visual inspection can be a quite powerful and often under-rated technique for detecting damage in composite structures. Even low-energy impacts may leave a slight marring, paint scrape, or faint surface blemish on a part. A slight wave or ripple on the surface may indicate an underlying delamination or disbond. A light spot or "whitish" area on a fiberglass part may indicate trapped air, a resin-lean area, or a delamination.

Figure 8-8. Visual inspection.

The "flat angle" or "grazing light" method of visual inspection involves examining parts at a very flat angle, towards a light source, and looking for the subtle shadows caused by minor waves from damage or defects.

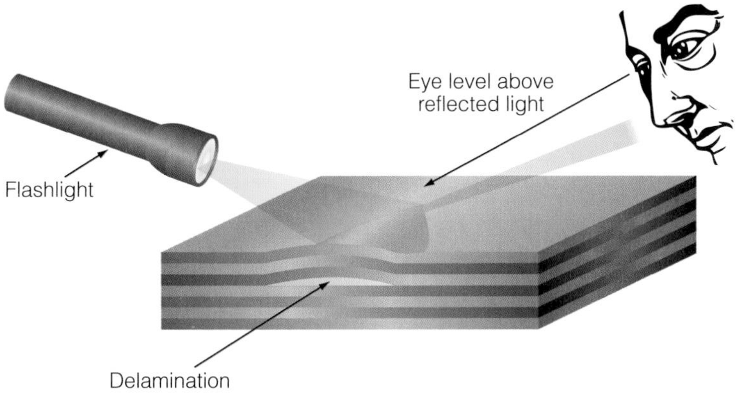

Eye level above reflected light

Flashlight

Delamination

Figure 8-9. Tap testing.

Using common tap hammer *(left)* and Wichitech RD3 digital tap hammer *(right)*.

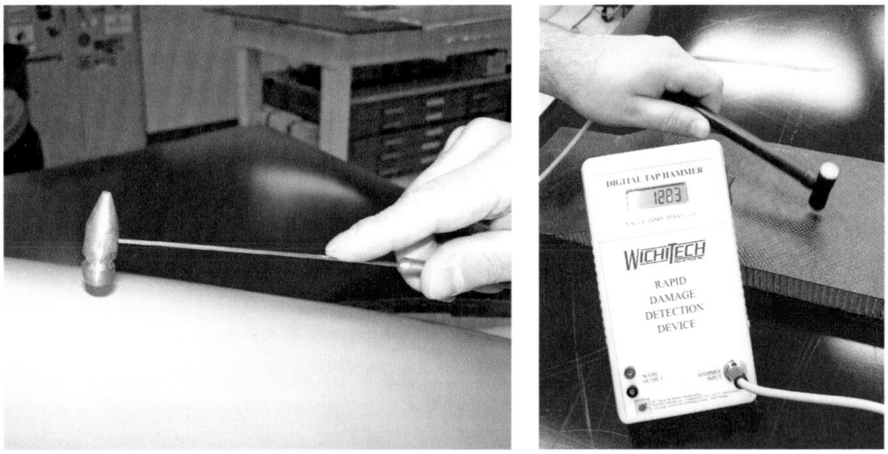

Tap Testing

Tap testing is probably the most common inspection technique other than visual. By tapping gently on the surface of a composite part, one can often hear a change in sound from a clear sharp tone to a dull thud. By tapping back and forth over the area in question, and making a small mark at the point where the tone just begins to change, it is possible to outline large, irregularly-shaped areas of delaminations or disbonds. However, there are many limitations to tap testing, including:

- Not good for deep damage.
- Requires knowledge of the underlying structural detail of the part.
- Not very effective in quantifying the degree of damage.
- Cannot locate small defects.

Ultrasonic Inspection

Ultrasonic inspection is used to detect flaws in a wide variety of materials, including metals and composites. It can be performed using portable battery-operated equipment, enabling parts to be inspected while still installed.

An ultrasonic testing (UT) instrument typically includes a pulser/receiver unit and a display device. The pulser/receiver unit includes a transducer probe that converts an electrical signal into a high frequency sound wave and then sends that wave into the structure being tested. A defect in the structure, such as a crack, causes a density change in the material and will reflect sound waves back to the transducer probe. The transducer converts the received sound waves (vibrations) into an electrical signal, which is then analyzed and shown on the UT display device. Inspection data may be displayed as an A-scan (time-based waveform display), B-scan (distance vs. time graph) or C-scan (plan view grayscale or color mapping).

A-scans are the typical display provided, but C-scans are becoming more common. Interpretation of these displays requires extensive training, and a skilled, certified operator is mandatory. B-scans are not normally used in

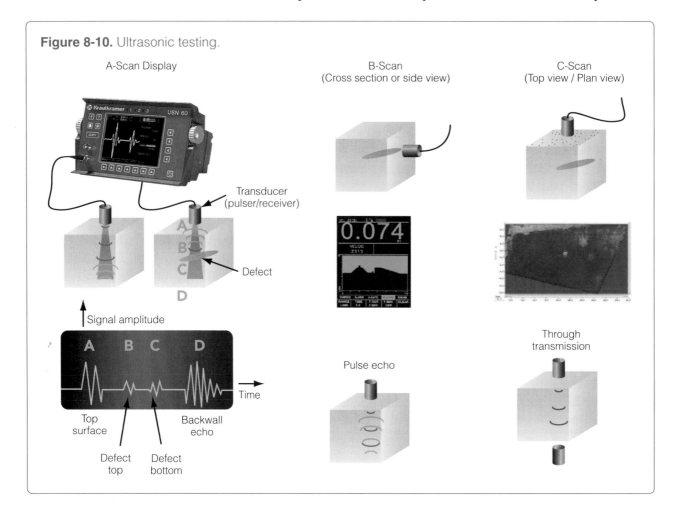

Figure 8-10. Ultrasonic testing.

composites because they look from the side, in-plane with the plies, and thus do not reveal as much useful information as A-scans and C-scans. UT equipment will allow, under favorable circumstances, measurement of not only the location but also the depth of the damage, which cannot be done visually or by tap testing. However, there are many limitations, including:

- Requires a couplant between the transducer probe and the part being inspected, as sound does not transmit well from air to solid materials. Couplant may be water or some type of gel.
- Calibration standards are required for each type of material and thickness.
- Interpretation of data demands careful analysis, requiring extensive skill, training and experience.

There are two basic techniques used in ultrasonic inspection: Pulse-echo and Through-transmission. Table 8.7 outlines the differences between these two methods.

TABLE 8.7 Differences Between Two Main UT Inspection Methods

Pulse-Echo	Through-Transmission
• One transducer generates and receives the sound wave.	• Uses two transducers: one to generate the sound wave and another to receive it.
• Usually limited to detecting first occurring defect—sound wave echoes back preventing detection of anything beyond.	• Defects located in the sound path between the two transducers will interrupt the sound transmission.
• More sensitive to misalignment between transducer and part (i.e. transducer should be within 2° of normal to the part).	• Interruptions are analyzed to determine size and location of defects.
• Better at detecting thin film inclusions in composites.	• Less sensitive to small defects than pulse-echo.
	• Cannot determine depth of defect.
• Can determine depth of defect by knowing speed of the sound wave through the composite.	• Better at detecting defects in multilayered structures and in quantifying porosity.
	• Can tolerate greater transducer misalignment with part (up to 10° from perpendicular).
	• Requires access to both sides of damaged part.

New UT Technology

Phased Array UT (*see* Figure 8-11 on the next page) is a relatively new offering, which may dramatically reduce the time required for inspection while providing excellent detection of small defects as well as the ability to determine both location and depth of defects. It achieves this by using an array of multiple transducers aligned in a single housing. By firing the transducers at slightly different times, the sound waves can be focused (depth) or steered (left, right, at an angle, etc.) toward a specific location. The transducer array is controlled electronically and can be programmed to sweep across and through the composite part. The transducer probe is also much larger than traditional UT probes, 4" versus 0.4", reducing the amount of time and movement across the surface required to provide 100% inspection of the part.

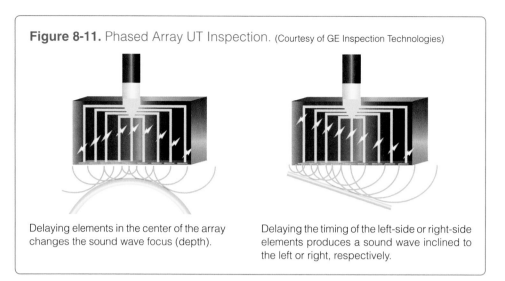

Figure 8-11. Phased Array UT Inspection. (Courtesy of GE Inspection Technologies)

Delaying elements in the center of the array changes the sound wave focus (depth).

Delaying the timing of the left-side or right-side elements produces a sound wave inclined to the left or right, respectively.

X-Rays

X-rays are a form of radiation from beyond the ultraviolet end of the spectrum. They are generated by directing a stream of electrons at a metal target, usually tungsten. X-rays pass through the test specimen, and then are sensed on the other side by a detection device. Traditional detection devices included film-coated plates or a fluorescing screen. Current state-of-the-art is a computer-controlled system comprised of a flat-panel detector and digital image processing unit.

X-ray inspection works better with fiberglass and aramid fiber composites than with carbon fiber. This is because carbon fiber and common matrix resins have very similar and very low x-ray absorption, resulting in an image that does not show much differentiation (i.e. appears black). X-ray inspection does work well in detecting transverse cracks, inclusions, honeycomb core damage and moisture ingression.

Because of the radiation involved, considerable safety protection is required when using this technique. Traditionally, the equipment was not very portable, but has become much more so over the past few years. The use of radio opaque penetrants helps to detect delamination and damage in complex structures, but also contaminates the part being inspected and eliminates the possibility of performing a bonded repair. In order to produce meaningful results, x-ray inspection requires a skilled, certified operator with significant training and experience.

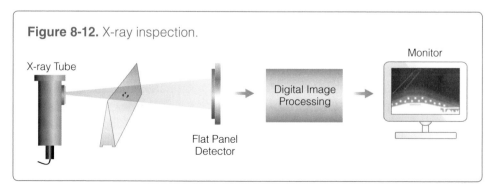

Figure 8-12. X-ray inspection.

X-ray Tube

Flat Panel Detector

Digital Image Processing

Monitor

Figure 8-13. X-ray.

X-ray inspection display of wing flap from an F-15 jet fighter, showing water trapped in honeycomb cells.

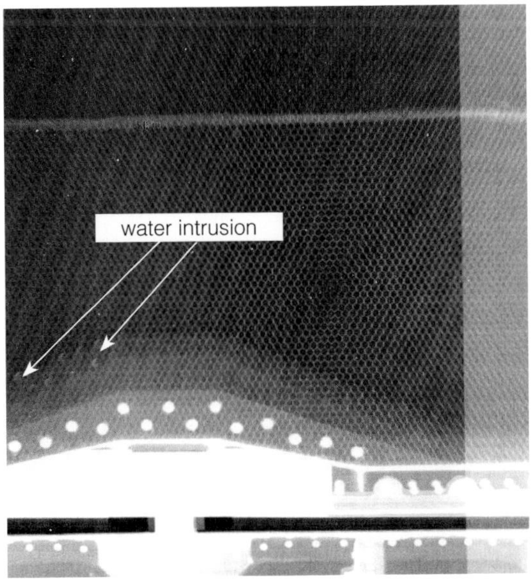

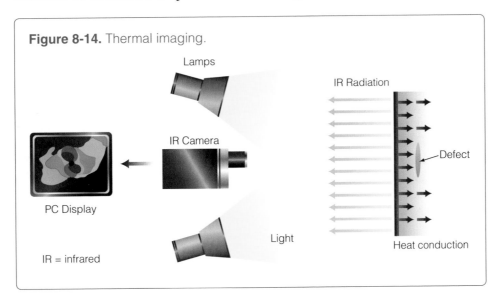

water intrusion

Thermal Imaging

This technique uses a heat source (typically a flash gun or heat lamps) to heat the composite part being inspected. As the part cools down, it is monitored with an infrared camera and digital processing equipment. Irregularities in the temperature distribution across the surface indicate the presence of defects.

Thermography offers a quick method for inspecting large areas and can provide a subsurface image of the entire structure. It requires fewer safety precautions than x-ray inspection and works well in detecting disbonds, delamination, inclusions, and variations in thickness and density. It is not as sensitive as ultrasonic inspection in detecting disbonds and delaminations,

Figure 8-14. Thermal imaging.

Lamps

IR Radiation

IR Camera

PC Display

Defect

Light

Heat conduction

IR = infrared

Figure 8-15. Thermal imaging vs. X-ray and UT inspection.

Three different inspection displays of the same panel having carbon fiber faceskins and Teflon inserts embedded in its aluminum honeycomb core. (Photos courtesy of Thermal Wave Imaging, Inc.)

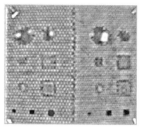
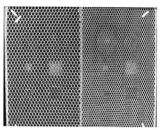
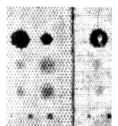

Flash thermography with thermographic signal reconstruction (TSR) processing

X-ray

Through-Transmission (UT)

Figure 8-16. Portable thermal imaging equipment.

Thermal imaging is now achievable with portable equipment that is easy to set up and break down. (Photos courtesy of Thermal Wave Imaging, Inc.)

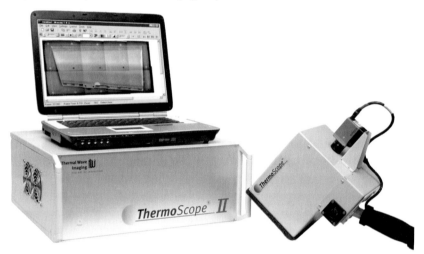

yet it does not require contact or couplants, can be performed with single-side access (i.e. part installed), and is now achievable with portable equipment that is easy to set up and break down.

Laser Shearography

Note: In this section, photography and text is provided courtesy of John Newman, Laser Technology, Inc., and used with permission.

Shearography nondestructive testing uses an image-shearing interferometer to detect and measure local out-of-plane deformation on the test part surface, as small as 3 nanometers, in response to a change in the applied engineered load. Shearography images tend to show only the local deformation on the target surface due to the presence of a surface defects subsurface flaws, delaminations, core damage, core splice joint separations, and impact damage.

Typical applied loads to the test part are dependant on the material reaction to the induced load. The optimum load type and magnitude depend on the flaw type depth, and is best determined before serial testing by making trial measurements. The applied load can be any of the following: heat, mechanical vibration, acoustic vibration, pressure and/or vacuum, electric fields, magnetic fields, microwave or mechanical load. Shearography offers exceptional high throughput and the ability to test a wide range of materials and structures including solid carbon fiber reinforced laminates, Composite Over-wrapped Pressure Vessels (COPVs), composite and metal honeycomb, foam cored sandwich structures and foam insulation of rocket launch vehicles.

Shearography is used both in production on new aerospace vehicles as well as during maintenance inspections. Applications include: AWACS rotodome, metal honeycomb control surfaces, radomes, air brakes, helicopter blades,

Figure 8-17. Shearograph of boron repair patch.

Performed on a C-141 aircraft shows small voids at center left and a large disbond at bottom right. (Photo courtesy of Laser Technology, Inc.)

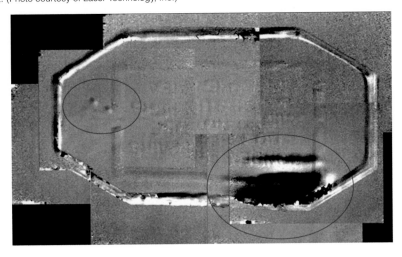

aluminum and titanium honeycomb panels and structures, foam cored structures and sandwich panels.

Portable shearography systems are designed for inspection of composite scarf joint repairs, boron repair patches, and non-visible impact damage on composite structures, particularly those with honeycomb core and thin facesheets. The entire system fits inside a transit case. The shearography inspection head vacuum-attaches in any orientation to the aircraft surface. Stored inspection procedures can be called-up to test specific parts that have been pre-programmed by an operator or Level III engineer.

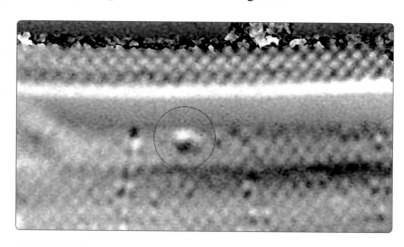

Figure 8-18. Laser shearograph of sandwich structure.

Shearography of titanium honeycomb and face sheet shows individual cells and disbond at center.
(Photo courtesy of Laser Technology, Inc.)

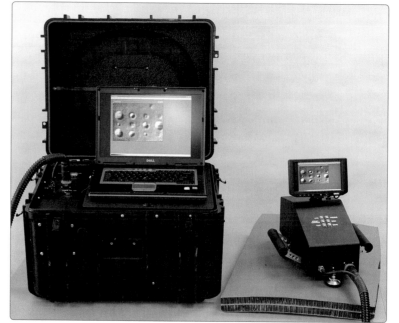

Figure 8-19. LTI-6200 portable thermal shearography system.
(Photo courtesy of Laser Technology, Inc.)

The production shearography system shown in Figures 8-22 and 8-23 is used at NASA Kennedy Space Center. The chamber includes an air blower capable of reducing internal pressure by 3–4 psi in five seconds and measures 8 x 8 x 12 feet. The internal shearography scanner envelope allows structures as large as 8 x 6 feet to be inspected automatically, stitching the images together to create a picture of the entire panel. This system is used to inspect composite honeycomb equipment for the Space Shuttle Orbiter.

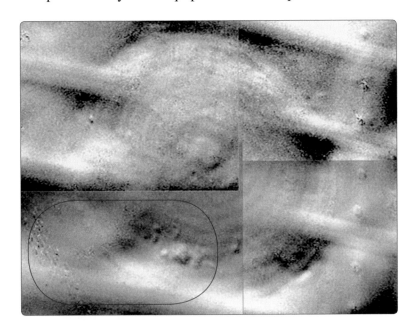

Figure 8-20.
Shearograph of engineered scarf joint repair.

Completed on an aircraft solid laminate with bonded stringers on the far side shows porosity as small pillowing bubbles at the bottom center and left center.
(Photo courtesy of Laser Technology, Inc.)

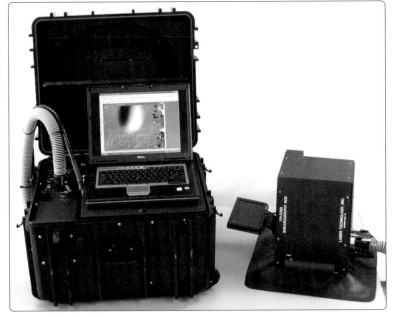

Figure 8-21.
LTI-5200 portable vacuum shearography systems, designed for large honeycomb and foam-cored structures.
(Photo courtesy of Laser Technology, Inc.)

Figure 8-22.
LTI-5200; testing a honeycomb panel on an RAF CH-47 helicopter.
(Photo courtesy of Laser Technology, Inc.)

Figure 8-23.
LTI-9000 production shearography system.
(Photo courtesy of Laser Technology, Inc.)

Holographic Laser Interferometry (HLI)

Note: In this section, illustrations and text are provided courtesy of Dr. Rikard Heslehurst, and used with permission.

Holographic interferometry is a non-contact optical NDI technique that provides visual representation of the out-of-plane deformations of an object. The out-of-plane deformations appear as bright and dark fringes superimposed on a three-dimensional image of the object being interrogated. Object interrogation is performed by loading the object in some form (mechanically, thermally or surface pressure) and exposing a holographic plate to two half-exposures of laser light. The two exposures are done at different load states. *See* Figure 8-24.

Holographic interferometry is very sensitive to surface movement. The differential height between two bright fringe lines is half the laser light wavelength. Using a He-Ne laser the image on the right shows height relief between two fringes of the order of 0.3 micro-meters (0.0003 mm or 0.0076 inch). Thus very small defects can be identified with holographic interferometry, such as weak bonding behavior. Holographic interferometry also shows how the

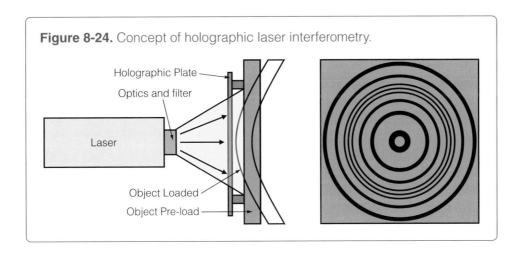

Figure 8-24. Concept of holographic laser interferometry.

Holographic Plate

Optics and filter

Laser

Object Loaded

Object Pre-load

surface is reacting to the presence of the defect that might be hidden within the object. For example, a holographic interferogram will show the behavior of a skin surface in a sandwich structure with damage on the other skin.

The holographic interferogram shown in Figure 8-25 was produced from initially placing a release film between the bondline of a thin sheet of aluminum and an aluminum plate. Note the fringe lines are uniform and further apart around the edges where the aluminum plates are well bonded. The center area depicts a disbonded area whereas the fringes are spaced closer together and in a less uniform manner. The partially bonded area exhibits fringes that are spaced randomly and indicate either a week bond or a partial disbond in this area.

(Both illustrations are from *Application and Interpretation of a Portable Holographic Interferometry Testing System in the Detection of Defects in Structural Materials,* by Heslehurst R.B., Ph.D., School of Aerospace and Mechanical Engineering, University College, University of New South Wales, 1998.)

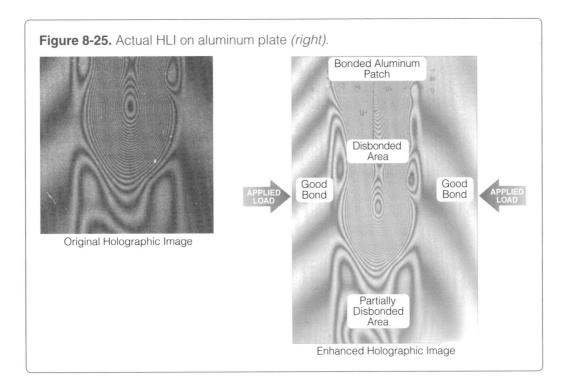

Figure 8-25. Actual HLI on aluminum plate *(right).*

Original Holographic Image

Bonded Aluminum Patch

Disbonded Area

APPLIED LOAD

Good Bond

Good Bond

APPLIED LOAD

Partially Disbonded Area

Enhanced Holographic Image

Dye Penetration

This technique is only mentioned because it is quite effective and well-known as a non-destructive method of inspecting metal parts for tiny hairline cracks. However, it is *not* to be used for composites. Although this technique will also show where cracks have propagated in composites, the dye penetrant liquid contaminates the composite part. By definition, the dye liquid is designed to wick itself into tiny cracks, which will contaminate any bondlines one might have in a repair scenario. Using this method eliminates any possibility of performing a bonded repair on the part. It can be useful on parts that will never be repaired, such as in a crash investigation, but it should not be used to assess the extent of damage in composite parts even though it is commonly used in metals.

Comparison of NDI Techniques

As shown in Table 8.8, each NDI method has its own strengths and weaknesses, and more than one method may be needed to produce the exact damage assessment required.

TABLE 8.8 Damage Inspection Methods

	Visual	Tap Test	A-Scan	C-Scan	X-Rays	Thermal Imaging	Holographic Laser Interferometry	Laser Shearography
Surface Delaminations	◖	●	◖	●	◖	●	●	●
Deep Delaminations	N/A	○	●	●	◖	◖	◖	◖
Full Disbond	◖	◖	●	●	◖	●	●	●
Kissing Disbond	N/A	○	○	○	N/A	N/A	●	◖
Core Damage	◖	◖	○	●	●	◖	●	◖
Inclusions	◖	◖	●	●	●	●	●	◖
Porosity	◖	N/A	◖	●	N/A	N/A	◖	N/A
Voids	◖	◖	◖	◖	◖	◖	●	◖
Backing Film	N/A	◖	◖	◖	◖	◖	◖	N/A
Edge Damage	●	◖	◖	●	●	◖	●	●
Heat Damage	◖	◖	◖	◖	N/A	◖	●	N/A
Severe Impact	●	●	●	●	○	●	●	●
Medium Impact	●	●	●	●	N/A	○	●	●
Minor Impact	○	○	○	○	N/A	○	●	●
Uneven Bondline	○	N/A	○	○	○	○	◖	N/A
Weak Bond	N/A	N/A	N/A	N/A	N/A	N/A	◖	N/A
Water in Core	N/A	◖	○	●	◖	●	N/A	N/A

● Excellent   ◖ Good   ○ Poor

Adhesive Bonding and Fastening

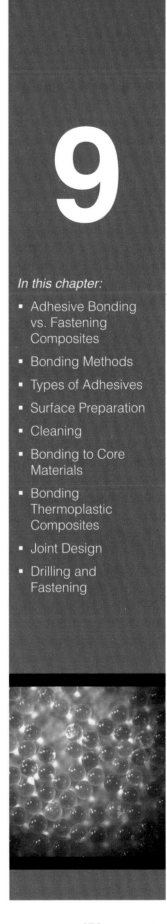

9

In this chapter:

- Adhesive Bonding vs. Fastening Composites
- Bonding Methods
- Types of Adhesives
- Surface Preparation
- Cleaning
- Bonding to Core Materials
- Bonding Thermoplastic Composites
- Joint Design
- Drilling and Fastening

Adhesive Bonding vs. Fastening Composites

In general, an adhesive bonded joint will transfer loads more efficiently than a fastened joint. In a well-bonded joint, the entire bonded area transfers loads between substrates. In a fastened joint, the loads are transferred primarily at each fastener location where the fasteners clamp the substrates together. Typically, a number of fasteners are required along the joint to accommodate these loads, and each through-fastener creates a hole in the composite laminate that weakens the structure at each location.

Long-term durability of an adhesively bonded joint is heavily dependent on many factors such as surface preparation, adhesive selection, timely application of the adhesive, uniform bondline thickness control at a targeted thickness, uniform clamping of the substrates, and proper cure of the adhesive. If any of these process factors are substandard, the adhesive bond will be compromised.

When designing for a fastened composite joint, it is important to build up the thickness of the locally fastened area to accommodate for the fastener grip length, countersink requirement (if applicable), and the bearing and shear strength of the joint. Fastener type and spacing as well as edge distance become the driving criteria for a fully functional, fastened joint.

It should be noted that thin laminates perform better when bonded than do thicker substrates. In general, very thick laminate joints should be considered for fastening and not for adhesive bonding. A combination of adhesive bonding and fastening may be an option in some joint designs. (Adhesive shim may be used at the interface.)

Bonding Methods

Co-Curing

Co-curing is the act of curing a composite laminate while simultaneously bonding it to some other uncured structure, or to a core material such as honeycomb, foam, or end-grain balsa. All matrix resins and/or adhesives are cured during a single process. Co-curing creates bonds that are more intimate than secondary bonds and it requires fewer cure cycles to produce a finished part. Co-cured structures generally are lighter than secondary-bonded structures. However, co-curing can be challenging with complex parts that have many details, which often require well designed, high-performance layup tooling.

Co-Bonding

Co-bonding is the curing together of two or more elements, where at least one has already been fully cured and at least one is uncured. An additional adhesive layer is usually required at the interface between the uncured and pre-cured materials. Co-bonding requires meticulous surface preparation on the pre-cured surface and often requires that one structure see more than one cure cycle. Co-bonding is commonly prescribed for structural repairs to composites.

Secondary Bonding

In secondary bonding, two or more already-cured composite parts are joined together by the process of adhesive bonding, during which the only chemical or thermal reaction is the curing of the adhesive itself. Secondary bonding requires meticulous surface preparation on the mating surfaces. It often requires precise tooling to control the location of the mating parts during the curing process.

Types of Adhesives

Liquid Adhesives

Liquid adhesives are usually lower in viscosity—less than 6,000 cP—and used in thinner bondlines, typically .002–.010 inch thick. They are often used with scrim cloth, non-woven mat, or microbeads to achieve thickness control. (See **Bondline Thickness Control Methods**, Page 176.) Liquid adhesives include room temperature and elevated temperature curing materials.

Paste Adhesives

These adhesives have a paste consistency, where viscosity is usually greater than 10,000 cP, and are used in slightly thicker bondlines, typically .005–.020 inch thick. They are also common in liquid shim/gap filling applications that are greater than .020 inches thick (this usually requires fasteners). Some paste adhesives may not wet out well on a cured composite substrate as surface-wetting may be inhibited by fillers in the formula. (See **Surface Preparation**, Page 176.)

Typically, scrim cloth, non-woven mat, or microbeads are used for thickness control. However, wetting of a fibrous material may be difficult with pastes. Microbeads may be a better alternative with thicker paste adhesives. Both room temperature and elevated temperature curing paste adhesives are available.

Film Adhesives

Film adhesives require frozen storage and out-time tracking similar to prepregs. The film thickness is precisely controlled by using a knit or non-woven mat as a carrier for the adhesive, with a range of .004 – .012 inch/0.10 – 0.30 mm thicknesses usually available. Note that these films are commonly specified not by thickness, but by aerial weight in pounds per square foot (PSF) or kilograms per square meter (Kg/m^2). Film adhesives are used for many core bonding applications and require elevated temperature cure.

Core Splice Adhesives

Core splice adhesives are not considered a structural adhesive but rather a core assembly adhesive. They are typically an expanding/foaming type of adhesive, increasing in volume by more than 200%. Core splice adhesives are used for bonding core assemblies together and for gap filling applications. They are typically supplied frozen in sheets, tapes and SemKit® (pre-measured, quick-mix) tubes and require elevated temperature and vacuum for proper processing.

Supported Film Adhesives

These film adhesives have a "carrier," usually of a very thin **non-woven mat**, woven scrim, or knit material to support the adhesive and aid in handling (*see* Figure 9-1). The carrier is also useful for maintaining bondline thickness control.

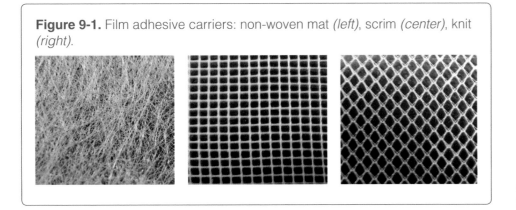

Figure 9-1. Film adhesive carriers: non-woven mat *(left)*, scrim *(center)*, knit *(right)*.

Unsupported Film Adhesives

These film adhesives do not have a carrier, and therefore will not control bondline thickness. One use for an unsupported film adhesive would be in a honeycomb sandwich structure where very low weight requirements might mandate a minimal areal weight requirement; or where the use of a carrier would otherwise not be desired within the laminate design. Unsupported film adhesives are traditionally hard to handle as they elongate and tear easily.

Reticulated Film Adhesives

These are typically unsupported film adhesives — designed to be heated, usually with a hot air gun, so that the hot adhesive runs to the edges and down into the cell walls of a honeycomb core before the face sheets are laid down. The cell walls become coated with adhesive, giving the desired fillets. Little extra adhesive is left over in the middle of the open cells after reticulation.

Another application is for acoustic panel bonding where the reticulated adhesive leaves open holes on the perforated skin, allowing for the attenuation of sound. Because there is no bond strength over the open spaces, a lighter adhesive may be used in some of these applications, saving weight while maintaining bond strength.

Bondline Thickness Control Methods

The ideal thickness of a cured adhesive bond depends on the type of adhesive selected. With a thin, low-viscosity adhesive, a thinner bondline is desired. With a thicker paste-adhesive, a thicker bondline may be required. Most high-strength structural aerospace-quality adhesives perform in a range from about .002 to .020 inches.

It is important to produce a bondline within the functional range of the adhesive and with a uniform thickness. Bondline thickness control is obtained with carriers in the adhesive as described below, or with the addition of very small, precisely controlled diameter **microbeads** in the adhesive mix.

Note that these are not the same as **microballoons**, which are not necessarily size-controlled. See Figure 9-2.

Microbeads are typically added to liquid adhesives in very small quantities, around 0.5% by weight. Some paste and liquid adhesives are available with the appropriately sized microbeads already mixed in by the adhesive manufacturer.

Surface Preparation

Metal vs. Composites

Surface preparation of metal surfaces often involve a series of chemical pre-treatments that are designed to first remove the weak oxidized outer layer

Figure 9-2. Microbeads vs. microballoons.

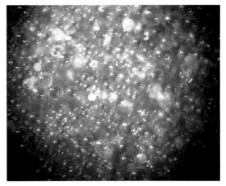

Microbeads are uniform in size having diametric tolerances as close as ±.0002 inches, thus allowing for precise B/L control.

Microballoons are non-uniform in size, but typically are within a specified range to obtain a desired mixing consistency.

of the metal substrate, then etch a highly mechanical micro-surface into the freshly exposed metal-oxide layer. This provides the maximum surface area possible on which to attach either an organic primer or an adhesive.

Surface preparation of (and ultimately bonding to) a fiber-reinforced plastic surface depends less upon mechanical attachment, and more upon a subtle chemical attachment precipitated by the attraction and eventual sharing of electrons at the interface between the adhesive and the substrate. This is known as a *covalent bond*. Achieving a covalent bond between the adhesive and the substrate surface is dependent upon the ability to raise the *surface free-energy value* (SFEV) of the composite surface to an equal or higher value than the (lowest potential) *surface tension value* (STV) of the adhesive. This electron exchange is crucial to achieving wet-out and a high performance chemical bond to a composite structure.

It may be noted that most metals, metal oxides, and ceramics exhibit a fairly high SFEV, $> 500 \text{ mJ/m}^2$. Most plastics do not, typically $< 50 \text{ mJ/m}^2$. For this reason, even the slightest compromise in the SFEV on a plastic surface can greatly affect the wetting on that surface.

Typically surface preparation to a fiber-reinforced plastic substrate is accomplished by lightly abrading with a suitable abrasive material, then cleaning the surface (without affecting the SFEV) so that it is free from dust or debris prior to bonding.

Mechanical Abrasion vs. Peel Ply

Achieving a suitable SFEV on the composite substrate is the goal, not mechanical roughness. One must shear up the top layer of molecules on the composite matrix surface, to create many broken atomic bonds without damaging or breaking the underlying fibers. Deep mechanical scratches from heavy abrasives might not only damage fibers, but also can be a route for moisture (or other fluid) ingress into the bondline over the life of the structure.

Therefore a peel ply is often used along the bonding surfaces of the composite substrate to keep the surface clean until ready for bonding and to provide for enhanced surface energy at the surface when removed. When it is removed, the peel ply fractures the resin matrix layer, leaving a fresh, slightly energized surface. This works well on smooth surfaces, however, on a woven cloth surface, peel ply removal may still leave small glossy depressions between the yarns in the weave pattern. These areas must be further prepared by careful abrasion for proper bonding.

Peel ply-fractured surfaces alone do not necessarily produce the highest SFEV nor the best bond-strength in lap-shear testing, but they usually provide consistent results. Many manufacturers prefer to use a peel ply-only preparation for bonding to get this consistency.

Release-Coated Peel Ply Fabrics vs. Non-Coated Peel Ply Fabrics

Release-coated peel ply fabrics will usually leave traces of release agent on the surface of the part, which may not be easily removed without damaging fibers. For this reason, non-coated fabrics are preferred for use along the bondline.

The problem with non-coated polyester or nylon fabrics is that they tend to stick well to the part laminate and are difficult to remove after processing. Fine-weave, prepregged polyester peel ply has been shown to work best under these circumstances.

Surface Abrasion Materials and Methods

The risks of damage associated with additional abrasion to the peel ply surfaces must be acknowledged and controlled. The use of common abrasion materials and techniques must be examined to determine what works best with the specific structures to be bonded. In addition, the potential for contamination from the abrasion material itself is a consideration when choosing a material or technique.

Surface abrasion can be accomplished with:

- ScotchBrite® Pads or disks (ultra-fine or fine)
- Fine grit sandpaper (#400 – #600 grit)
- Abrasive media blasting (typically #18 – #220 mesh) alumina grit

Hand-abrading with ScotchBrite® type abrasives are widely recognized as the preferred material of choice for pre-bond preparation of composite materials and present the least risk of fiber damage, while abrading with sandpaper or performing any type of media blasting presents a greater risk for damage. The risk of contamination with melted nylon from the ScotchBrite® type disks is present when using high-speed grinders or sanders, while contamination from the binders or non-clogging agents on sandpaper is also a risk.

Because plastic surfaces are generally considered to be low energy surfaces (< 50 mJ/m^2), it is necessary to select a compatible adhesive that has a suitable surface tension value for achieving wet-out on the energized plastic surface. It is also important to note that the "raised energy" condition, once achieved, is short-lived: within an hour or so, the effect is lost. Thus surface preparation and bonding should be accomplished in a timely manner.

Water Break Testing*

A *water break* test can be used to determine the surface energy change on a *sample* piece of substrate.

*(Note: This test is *not* recommended for use on an *actual* composite substrate surface to be bonded!)

If the surface energy produced is high enough, clean distilled or deionized water (STV approximately 70 mJ/m^2) will spread out in a slightly arched film on the surface (e.g., not a completely flat sheet), and should not "break" into beads.

If this type of water break-free surface can be maintained for 20-30 seconds, then you have achieved a clean, high-energy surface suitable for adhesive wetting. Subsequently, the same surface preparation process can be followed on the actual substrate. If the surface is contaminated by the abrasive, or exhibits a much lower surface-free energy value than the water, the water will break into rivulets and bead up on the surface. This surface preparation would probably yield a less than adequate surface preparation for bonding.

Cleaning

Solvent Cleaning

Many specifications still require solvent wiping as part of the process. If solvent cleaning is specified, than the double-wipe method using a reagent grade solvent would most likely be the method of choice. The double-wipe technique uses one wet cloth followed immediately by one dry cloth to absorb the solvent before it can evaporate.

Solvent Grades

Commercial grade: Inexpensive, but has an undocumented contaminant level. Normally this is suitable for shop use. (Not recommended for cleaning prior to bonding.)

Reagent grade: Expensive, but certified as clean. Typically contaminants are listed in parts per million (PPM) in the certification documents. Strongly recommended if solvent is to be used in a pre-bond cleaning operation.

Common types of solvent, in order of preference:

- Acetone
- Methyl Ethyl Ketone (MEK)
- Methyl Isobutyl Ketone (MIBK)
- Isopropyl Alcohol (ISP)

Non-Solvent Wiping Technique

A non-solvent wipe alternative has been recently recognized as a possible preferred technique for cleaning composite surfaces prior to bonding. Careful vacuuming and/or dry wiping with approved wipes to remove surface debris is recommended. Multiple passes may be required to remove the dust and debris to a suitable level.

Since no solvent is used, health, safety and waste disposal problems are minimized. In addition, there is no risk of creating porosity in the adhesive bondline due to residual solvent flashing to steam during high-temperature cures. Some test results indicate bond strengths and longevity using the dry-wipe method are equal to, or better than those using solvent wiping.

Wipes

Pre-bond wiping should always be done with an approved, non-contaminating wipe. Industrial woven-fabric rags and cloths may have chemicals or *size* in the material that can be re-distributed on the freshly prepared bond-surface. This is especially true when using solvents that might break down these chemicals into solution.

Lint-free cloth exists that might still have a "size." Also, size-free cloth exists that may leave lint or traces of fibers on the surface when used. The latter might be of lesser concern because the chemical bond would not necessarily be affected by trace amounts of lint or fiber, whereas it may be drastically affected by the presence of incompatible chemicals.

Bonding to Core Materials

Bonding to Honeycomb Core

Honeycomb core bonding is best accomplished with a film adhesive that is specifically designed for such uses. This may be an unsupported film adhesive, since bondline thickness control is not required. The essential ingredient for a good honeycomb core bond is the formation of fillets of adhesive down the sides of the core cells. These fillets give the bond a large surface area, and the adhesive is subsequently loaded in shear, which is the most desirable loading mode. (*See* Figure 9-3.)

Bonding to Foam Core

Bonding to foam core is very different than bonding to honeycomb. A good bond to a foam core depends on having a suitable amount of adhesive to wet the porous foam surface and actively (both chemically and mechanically) attach to the material. Often, the adhesive bond to a foam core is much stronger than the core material itself. Frequently, failures to foam core sandwich panels occur in the foam along a line through the mean depth of the surface

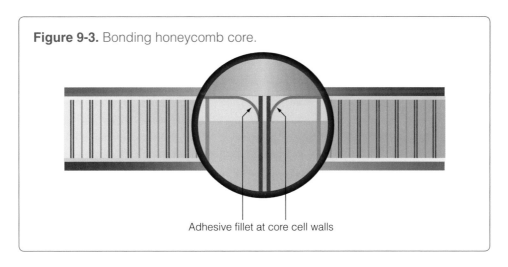

Figure 9-3. Bonding honeycomb core.

Adhesive fillet at core cell walls

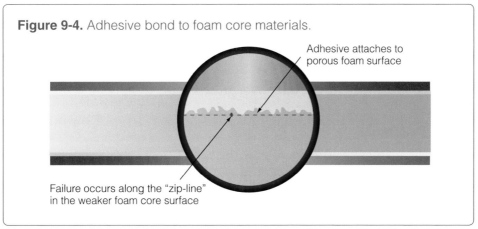

Figure 9-4. Adhesive bond to foam core materials.

Adhesive attaches to porous foam surface

Failure occurs along the "zip-line" in the weaker foam core surface

pores of the material. This "zip-line" failure effect is well recognized by those who work with foam core sandwich structures. (*See* Figure 9-4.)

Bonding Thermoplastic Composites

Most thermoplastic polymers are not easily bonded using adhesives common in composites, such as epoxies, due to inherent characteristics of low surface energy and non-polar surfaces. These difficulties can be resolved through surface treatments or the use of fusion bonding techniques.

Surface Treatments

There are a variety of surface treatments used to increase the wettability and polar nature of thermoplastics, including: plasma treatment, corona treatment, flame treatment, and acid/chemical etching. Although these treatments can be effective, they also have several disadvantages. The treated surfaces maintain their characteristics for only a short time, and thus require adhesives to be applied relatively quickly and, in the meantime, surfaces must be protected from contamination. The equipment required is typically expensive and the maximum part size which can be treated might be restricted.

The use of high voltage and/or flame also presents significant hazards and might not be practical for certain manufacturing environments and field repair situations.

Plasma Treatment

Plasma treatment takes a gas such as oxygen, argon, helium, or air, and excites it by applying a high voltage between electrodes in a low-pressure chamber. The substrate surface is then bombarded with the generated ions, forming reactive groups which increase its reactivity and wettability.

Corona Treatment

Similar to plasma treatment, corona treatment (also called corona discharge) involves creating a plasma in air at atmospheric pressure near the substrate surface by exposing it to a high-voltage electrical discharge. The ionized air appears as a blue-purple glow with a faint sparking—a corona—and results in the generation of reactive groups on the surface that serve as potential bonding sites for the adhesive, greatly improving bond strength.

Flame Treatment

In flame treatment, a combustion flame is aimed at the substrate surface, oxidizing it and creating polar groups, which create a surface that is more wettable by the adhesive.

Acid/Chemical Etching

This surface treatment uses chemicals such as chromic or nitric acid to provide a combination of surface oxidation and microscopic roughening.

Crystalline vs. Amorphous Thermoplastics

Airbus, Ten Cate, Stork Fokker and other companies have found that amorphous thermoplastics such as PEI can be satisfactorily adhesively bonded using conventional adhesive systems.

Crystalline thermoplastics, such as PPS and PEEK, present more challenges. Studies performed in the late 1980s and early 1990s showed that the use of FM300 film adhesive (a common epoxy system used in composites) combined with a plasma treated surface gave the greatest lap shear strength of 41.6 MPa for AS4 carbon fiber/PEEK laminates. Grit blasting and acid etching produced lap shear strength values of only 20 MPa.

However, Airbus has found that even with sufficient surface treatment, there are still issues with hot/wet performance of adhesive bonds with crystalline thermoplastic composites. Airbus investigated the use of adhesive bonding with PPS structures, using a variety of surface treatments. These included alumina blasting, corona discharge, flame treatment and adhesion promoters, and also a variety of adhesives including two-part epoxies, two-part urethanes, and epoxy film adhesives. They found that none of the adhesives and/or surface treatments produced results which met their minimum hot/wet lap shear requirements of 1,000 psi. Thus, Airbus has not approved any adhesive bonding techniques for manufacture or repair of crystalline or semi-crystalline thermoplastics such as PPS and PEEK.

Fusion Bonding

Fusion bonding involves heating and melting the thermoplastic polymer on the bond surfaces of the components and then pressing these surfaces together for consolidation and solidification. Unlike an adhesive bond, which under a microscope retains a clearly defined joint line formed by the adhesive film, materials that have been fusion bonded (i.e., welded) thoroughly intermix or fuse through, becoming essentially a single part.

Fusion bonding has several additional advantages over adhesive bonding, such as reduced cycle time, less surface preparation, no shelf life or bonding immediacy issues, and the absence of stress concentrations. However, the heating must be controlled and contained so that the plastic structure of the components being bonded is not significantly altered, and the parts must be kept under pressure to prevent warping. There are four basic types of fusion bonding used with thermoplastic composites.

Thermal Welding

In this method, heat is directly applied to the surfaces being joined. The heat is generated from an external source such as a hot plate, flame, lasers or infrared heater.

Ultrasonic Welding

Ultrasonic welding uses high frequency sound energy to soften or melt the thermoplastic surfaces at the joint. The parts to be joined are held together under pressure between the oscillating horn and an immobile anvil, then are subjected to ultrasonic vibrations at a frequency of 20 to 40 kHz at right angles to the contact area. Alternating high-frequency stresses generate heat at the joint interface to produce a good quality weld.

Ultrasonic welding is an easily automated and very fast process, with weld times of less than 1 second. However, because the equipment for this process are expensive, it has traditionally been used for large volume production runs in the plastics industry, and typically limited to small components with weld lengths not exceeding a few centimeters.

Induction Welding

Induction and resistance welding are the most commonly used fusion bonding methods for thermoplastics, and considered to be the most promising. Both make use of an insert which helps to generate electrical current that then heats the thermoplastic substrates. After cooling down under pressure, a bond is created. In induction welding, the insert is called a susceptor, conductor, or conductive element; in resistance welding it is called a resistive element or welding tape. In both processes, it is usually some type of wire mesh.

Induction welding (also called electromagnetic welding) uses a high radio frequency generator to create an electromagnetic field, which then produces an alternating electric current. The current passes through the conductive element, which heats up, softening and fusing the thermoplastic substrates. The whole process takes place within specially designed fixtures that join the parts together under low pressures and takes only seconds for a typical weld.

Induction welding has been used for 30 years in the plastics industry, mostly with polyolefins. Because the electromagnetic energy precisely targets the susceptor/conductor, induction welding doesn't significantly affect the dimensional stability of the mating components. It also does not distort the material adjacent to the bondline, as is common with welding based on conductive heating or the application of mechanical energy to the joint. Induction welding can easily handle large parts and can also support a fair amount of automation; for example, the susceptor/ conductor implants can be installed robotically or even molded into the mating parts.

Resistance Welding

In resistance welding, electric current is passed directly into the insert, which is an electrically resistive element—usually a strip of metal mesh, also called welding tape. Specially made tooling clamps the welding tape between the two parts at the intended bondline. The metal mesh tape quickly heats as electric current is applied, heating the adjacent part surfaces to the melt temperature of the thermoplastic matrix. The mesh—which is typically very thin (~0.2 mm/0.01 inch thick) and open—remains in the joint without disrupting the bond. As with induction welding, heat is introduced into the structure only where it is required, and thus limits the material melt zone, reducing the risk that part shape or dimensional stability will be compromised. The total time to weld is typically a couple of minutes for parts ranging in length from 3 to 4 meters (9.84 to 13.1 feet); therefore, the process adapts well to production and can be automated. It is currently being used by Stork Fokker to assemble the thermoplastic composite J-nose leading edge assemblies used in the Airbus A340 and A380 commercial jetliners.

Thermabond Process or Dual Resin Bonding

The process of induction and resistance heating using an amorphous thermoplastic film or prepreg has been evaluated by the Australian Department of Defense to be one of the best methods for in-field composite repair, if an appropriate thermoplastic film resistant to chemical attack and environmental degradation is used. Patented by ICI-Fiberite in the 1980s as the Thermabond process, it does not require complex equipment and is not unlike the adhesive bonding repair techniques currently used by most military aircraft repair operations. The successful and repeatable use of acoustic emission as a quality control procedure is a further positive for the use of this bonding method with thermoplastic composite structures. An example from a manufacturing setting is shown in Figure 9-5, where two PEEK composite laminates are molded with an additional layer of lower melting point PEI film on their surfaces. These surfaces are then joined by placing one more ply of the PEI film or prepreg in between, and applying local heating to melt the film but not the substrates.

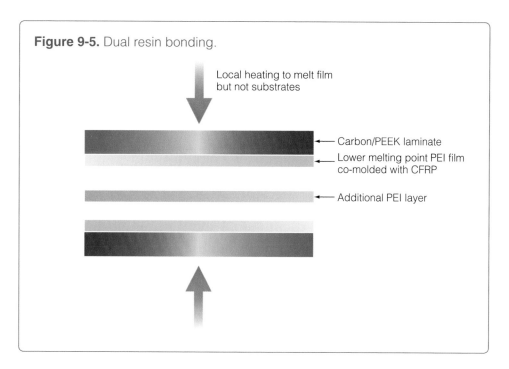

Figure 9-5. Dual resin bonding.

Local heating to melt film but not substrates

Carbon/PEEK laminate

Lower melting point PEI film co-molded with CFRP

Additional PEI layer

Joint Design

Many different configurations of adhesively bonded joints are possible. (*See* Figure 9-6 on the next page.) The "scarf" joints, with flat scarf angles, are generally considered to have the best strength. However, flat scarf angles can be difficult to machine, and achieving proper fit on pre-cured parts being secondarily bonded can be almost impossible using hand techniques. However, co-cured repairs can be done easily and successfully.

Drilling and Fastening

Mechanical Fastening of Composites

Mechanical fasteners have long been employed to join materials and to build complex multi-piece structures. Fasteners have evolved over time to meet the specific needs of a variety of different materials and applications. As a result, fastener designers have engineered an assortment of fasteners that work best with advanced composite materials.

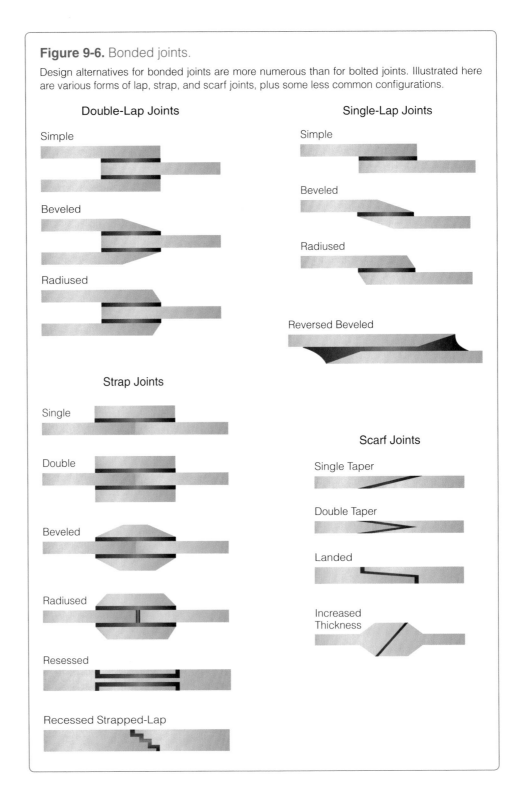

Figure 9-6. Bonded joints.

Design alternatives for bonded joints are more numerous than for bolted joints. Illustrated here are various forms of lap, strap, and scarf joints, plus some less common configurations.

Double-Lap Joints

Simple

Beveled

Radiused

Single-Lap Joints

Simple

Beveled

Radiused

Reversed Beveled

Strap Joints

Single

Double

Beveled

Radiused

Resessed

Recessed Strapped-Lap

Scarf Joints

Single Taper

Double Taper

Landed

Increased Thickness

Bonding of Thin Composites

In general, bonding thin composite structures is preferred over mechanical fastening, as the loads are distributed more uniformly across the joint. While bonding is usually desired, it is not always practical wherever access to the internal structure or panel-removal/replacement may be required. Often, extra plies are needed in local areas of the adjoining panels in order to accommodate bearing loads and fastener grip-length requirements.

Conventional riveting, with a rivet gun and bucking bar, is never used in composite structures due to the assured delamination that will occur during the installation of these fasteners. To reduce the risk of damage to laminated structures during installation, many traditionally designed blind rivets, blind bolts, and lock bolts have been modified to accommodate the peculiarities of advanced composite materials. In addition, specially-designed large "footprint" fasteners are used to more evenly distribute the loads into the laminated panels, without inducing concentrated loads at the edges of the head and foot of the fastener.

Mechanical Fastening of Thick Composites

In thick, highly loaded composite structures, fastening is preferred over bonding because the loads through the joints may exceed the capabilities of common structural adhesives. In these cases, large diameter, high-bearing-strength fasteners are used, often requiring an interference fit in the laminated panels. This is a problem when using conventional fasteners and techniques. Driving an oversized fastener through a composite panel is not an option; the panel will certainly delaminate around the hole and reduce the strength of the surrounding structure. Because of this, special large-head, expanding-sleeved fasteners are prescribed for these applications to lessen the chance of damage during installation.

Several manufacturers have successfully developed fasteners that deliver an interference fit without damaging the laminate, most notably Alcoa's Huck-Tite® fastener (see Figure 9-7). This fastener uses a stainless steel sleeve which expands diametrically inside the hole as the titanium fastener body is drawn through it, resulting in an extraordinarily tight fit that resists fatigue and fluid ingress.

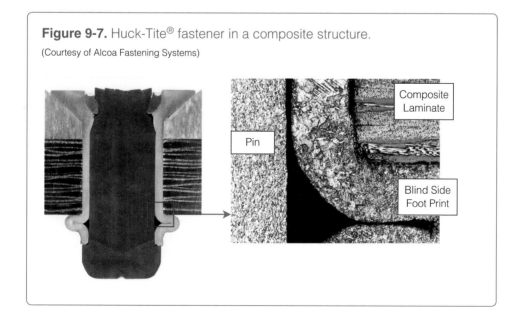

Figure 9-7. Huck-Tite® fastener in a composite structure.
(Courtesy of Alcoa Fastening Systems)

It is found that full bearing strength is developed when the edge distance ratio, e/D, is equal to or greater than 3.0 and the side distance ratio, w/D, is equal to or greater than 4.0.

Fastener Footprint

Mechanical fasteners traditionally used in metallic structures have a relatively small footprint. That is, the contact area on both sides of the fastened substrates is minimal. While this design works fine on metallic structures, it is wholly inadequate for structures made from advanced composite materials. When a mechanically fastened joint is loaded in shear, the upper half of the fastener is pulled in one direction, and the lower half is pulled in the other. This action causes the fastener to "tip" in the hole.

Older fastener designs with smaller heads did not have sufficient leverage to resist this tipping action in composite structures and would therefore tend to give way and dig into the fibers at the edges of the fastener head. This combination of factors led to premature fatigue in the fastened composite joint. Today, fasteners designed for use in composite structures have a much larger footprint to help them combat the tipping forces exerted on them when loaded in shear.

Figure 9-8. Large and small fastener footprint area.

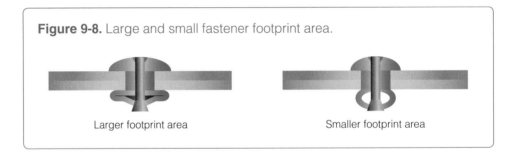

Larger footprint area   Smaller footprint area

Fastener Materials

Many metals such as aluminum alloys and many types of steel that are traditionally used to make mechanical fasteners will corrode when placed in contact with carbon fiber composites. As is the case with dissimilar metals, galvanic corrosion occurs because carbon is at the opposite end of the corrosion scale from many of these standard materials. (*See* the galvanic series chart in Table 4.3, Page 84.)

This is not the case with glass or aramid fiber composites as they do not conduct electricity. However, fasteners designed for use in composite structures are usually made from more noble metals that are compatible with carbon applications. Titanium is the first choice for fasteners due to its exceptional compatibility with carbon fiber. Titanium also has an unmatched strength-to-weight ratio and can tolerate extreme temperatures, but is very expensive. A less costly alternative is stainless steel. A-282 and A-286 stainless steel are the preferred alloys for fasteners when titanium is either cost-prohibitive, or the extra weight of the stainless fasteners is deemed acceptable.

Hole Tolerances

Another contributing factor to joint fatigue is the quality of the installation hole. The hole must be normal to the surface and must be the proper diameter. This is largely due to the fact that, unlike their metal counterparts, composite laminates are rather fragile edge-on, and are unable to handle edge loading inside the fastener hole unless the fastener fits tightly during installation, and remains so throughout its service life.

A net-diameter hole is somewhat difficult to achieve in high fiber volume composite structures due to the nature of the materials involved; the ends of the fibers tend to move away from the drill or ream during machining and then "rebound" back into place afterwards. This results in slightly smaller diameter holes and a tendency for the fastener not to fit. Because of this, the importance of proper tool selection and the use of proper techniques cannot be overstated.

Drilling Tools and Techniques

The advanced composites industry is rife with special tools to cut, machine, and drill composite materials. Equally impressive is the voluminous quantity of text concerning techniques for drilling and machining composites that has been published over the last few decades. This is partly due to the fact that each fiber type has its own set of unique machining characteristics, and these characteristics may change, perhaps dramatically, depending on the volume ratio and properties of the matrix in which they are embedded.

Speed and Feed Rate

For years, the rule of thumb in machining composites has been *high speed–low feed*. In other words, the tool's cutting edge(s) should be traveling as fast as is practical for the operation, and the operator should allow the tool time to do its job by not using excessive feed rates.

At lower RPM, even drill bits especially designed for use in composites have a tendency to generate more delamination at the surface plies, as well as on the far side of the laminate as the drill exits the work. In general, only high RPM drill-motors capable of turning at between 5,000 and 22,000 RPM, with sufficient torque to maintain these speeds, should be used for drilling composites.

Feed rates vary tremendously with the type of laminate being drilled; however it is important to never force the drill to cut faster than its design or condition allow. This will usually translate into excess heat within the laminate and may compromise the matrix material. Assuming the proper tool has been selected for the material being drilled, common feed rates vary from 15 up to 60 inches per minute.

The goal is always to drill the best hole possible in the least amount of time. In doing this, there is always the danger of developing excessive heat within the laminate. Often, to reduce this danger, lubricants are suggested or even specified for some drilling operations. Lubricants especially formulated for use in composites also have the added benefit of promoting longer tool life, and allow operators to push the upper limits of the recommended feed rate for any given material. While this will certainly speed-up drilling operations, one must assess whether or not any residual lubricant (i.e., contaminant) will hinder subsequent operations such as adhesive bonding, sealant application, or painting.

Controlling Angle and Feed Rates

As mentioned earlier, successful installation of fasteners in composite materials relies on strict adherence to the manufacturer's specifications and tolerances. Chief among these are the ones concerning installation holes. Not only must the holes be of the correct diameter, but they must be perpendicular to the surface being drilled.

Delamination on the far side of a hole is probably the most common difficulty when drilling. Drill design, speed and firmly backing up the work are important, but sometimes more is required to get satisfactory results. Often, a controlled-feed drill will provide the extra level of control necessary. As the name suggests, a controlled-feed drill restricts the speed at which a drill bit penetrates the work. This eliminates any tendency a bit might have to "screw" its way through the laminate and generate delaminations. Some manufacturers, like Monogram, offer a kit with a controlled-feed drill motor and a fixture designed to ensure that the drill remains normal to the surface throughout the drilling process (*see* Figure 9-9 below). In other cases it might be necessary to use a conventional drill guide such as the one shown in Figure 9-10.

Figure 9-9.
Monogram controlled feed drill and fixture.

(Photo courtesy of Monogram Aerospace)

Figure 9-10. Conventional drill guide—uses a slip-renewable bushing *(left)* in a pressed bushing in a drill fixture or template.

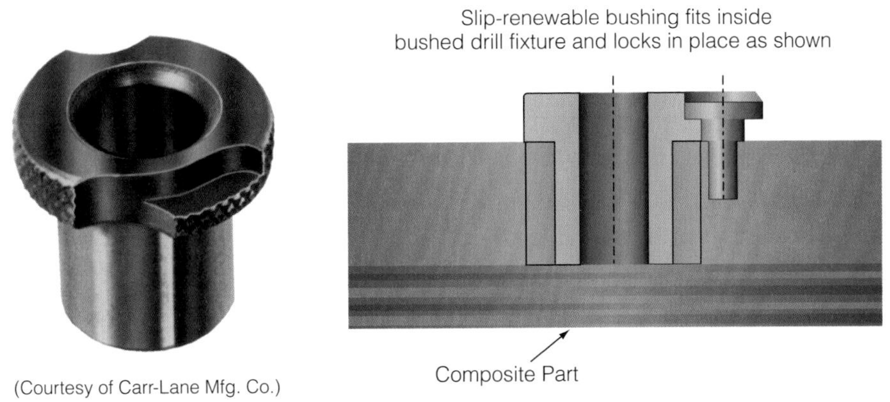

Slip-renewable bushing fits inside
bushed drill fixture and locks in place as shown

(Courtesy of Carr-Lane Mfg. Co.)

Composite Part

Drilling Carbon and Glass Fiber Reinforced Composites

Carbon fibers are relatively stiff and brittle, while glass fibers are more flexible by comparison. Despite this, they have a lot in common when it comes to drilling and machining. Thus, many tools that are designed to cut, machine, or drill carbon fiber composites also work well on fiberglass laminates.

Carbon and glass laminates are extremely tough on cutting tools, able to quickly round off the cutting edge on any high-speed steel (HSS) tool. The solution here is to manufacture the drill (or countersink) from a material hard enough to maintain a sufficiently sharp edge against carbon and glass. Two such materials are carbide and diamond.

Carbide drills designed for use in composites usually have a positive (often extreme) rake angle, a fine chisel edge, and a gradual taper up to the final drill diameter. An acute positive rake angle encourages the fibers to be drawn into the cutting area and sheared between the cutting edge and the uncut material. A positive rake removes more material in a shorter amount of time than a negative rake, but it also makes the cutting edge more sensitive and fragile. The penetration rate, and thus the production rate, of a drill bit can be enhanced by incorporating a small chisel edge at the point. Because of the proportionately high-penetration rate, the optimum chisel edge for drilling composites would be as close to a sharp point as possible.

These features conveniently harmonize with a gradual taper in the bit that is often incorporated into drill designs. The gradual taper minimizes cutting pressures at the edge of the hole, gradually opening it up to its finished dimension. Additionally, this feature helps to minimize breakout and delamination on the far side of the laminate that can occur with standard twist-drill configurations. Regardless of the drill design, supporting the backside of the composite panel with wood or suitable fixturing is recommended to minimize backside-fiber breakout.

Diamond coated carbide tools are highly effective for cutting and drilling carbon fiber and fiberglass structures. With these tools, small diamonds of a specific grit size are deposited on the cutting edge and are embedded in a metal matrix material. Ideally, as the sharp edges of the exposed diamonds are worn away, new diamond edges begin to appear as the matrix is worn away. This provides additional longevity of the cutting tool. Diamond coating technology is extremely versatile and can be used for drill bits, hole saws, cutting wheels, countersinks, and a myriad of other composite machining applications.

Drilling Aramid Fiber Reinforced Composites

Aramid fiber's incredibly high tensile strength and durability has been leveraged for years by engineers looking to design abrasion-resistant, lightweight structures. Unfortunately, that very same toughness and durability become liabilities when machining aramid composites.

Conventional twist drills tend to produce rough, fuzzy edges in aramid fiber laminates that might be considered unacceptable in most applications. This is because these drills tend to grab the fibers and pull them in tension until the fibers break, rather than shearing them neatly. While the resin content of all composite laminates affect the machinability, this is especially true of aramid reinforced laminates. Resin-rich laminates tend to make for cleaner holes, while resin-lean laminates tend to have more fuzz at the edges.

One of the better drilling solutions for aramid composites is the Brad-point carbide twist drill. This self-centering, modified, two-flute twist drill features a sharp center-point surrounded by two sharp peripheral cutting edges that draw the fibers in tension and then shear them off. This results in a nearly fuzz-free hole through most aramid laminates.

DuPont, who makes Kevlar® aramid fiber, recommends the use of Brad-point drills at a speed between 6,000 and 25,000 RPM and with a slow, controlled feed rate. They also suggest that, unlike other composites, RPM should be decreased as the size of the drill bit increases.

Figure 9-11. Brad-point drill for drilling aramid fiber laminates.

(Courtesy of Traditional Woodworker)

Other Approaches

As we've seen, using mechanical fasteners in advanced composite materials poses some unique problems. Unlike metal structures, composites are tremendously strong and lightweight, yet can be fragile and easily damaged. These peculiarities must be considered when considering what type of fasteners should be used in a given composite structure. When blind rivets/bolts or lock bolts are impractical or simply not appropriate to the application, potted inserts or adhesively bonded fasteners may be the solution to the fastening problem.

Inserts

The need for an effective means to attach components to honeycomb sandwich structures was realized shortly after the first such panel was built. The relatively delicate, light-weight honeycomb core material adhesively bonded to two thin skins was ill-suited to accommodate significant loads in any direction. For decades, potted inserts provided the capability to attach components and sub-assemblies to such sandwich structures. Rosán Delron Inserts offers a wide choice of types, styles, materials and finishes for sandwich panel fastening.

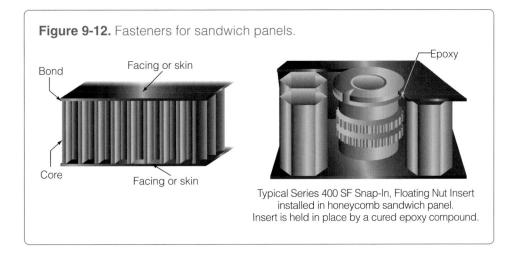

Figure 9-12. Fasteners for sandwich panels.

Bond

Facing or skin

Epoxy

Core

Facing or skin

Typical Series 400 SF Snap-In, Floating Nut Insert installed in honeycomb sandwich panel. Insert is held in place by a cured epoxy compound.

Bonded Fasteners

In cases where threaded studs, sleeves, or other specific hardware must be installed, adhesively bonded fasteners may be the best option. Click Bond, Inc. has developed numerous solutions to these fastening challenges for use on advanced composite structures. Adhesive bonding of fasteners and/or other hardware has a number of advantages over previous installation methods. For example, installing an adhesively bonded nut-plate requires that only one hole be drilled in the structure since the need for the retention rivets is eliminated. Additionally, it takes approximately 80% less time to install these fasteners versus riveting them on.

Figure 9-13. Adhesively bonded fasteners.

(Courtesy of Click Bond, Inc.)

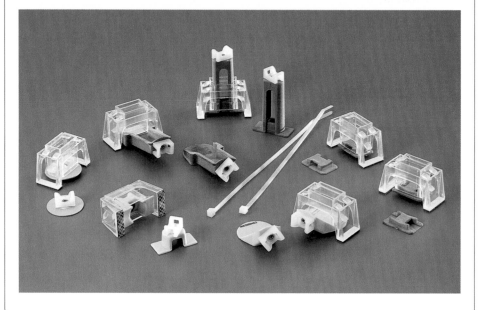

Repair of Composite Structures

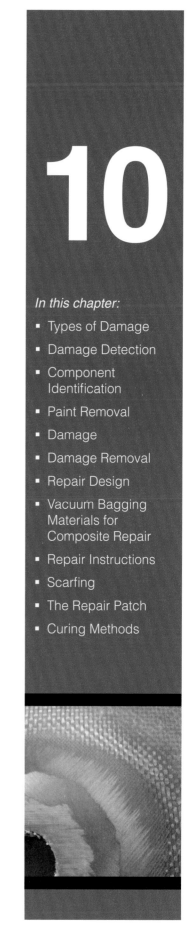

10

Types of Damage

Most composites are damaged while in-service from overloading, impact or environmental effects. Parts may also need to be repaired due to a manufacturing defect. Such defects may make a composite structure more susceptible to in-service damage.

Holes and Punctures

Holes and punctures are usually caused by high or medium level impacts. They can be severe, but are usually easily detected. (*See* Figure 10-1 on the following page.)

Delaminations

Perhaps the most common of all damage types, delaminations are often caused by low-energy impacts, such as a tool drop or a glancing bird strike on an aircraft. They can sometimes be visible if they are near the surface. Delamination can also be caused by manufacturing defects, resin starvation in a laminate and by **moisture ingression**. (*See* Figure 10-2 on the following page.)

Disbonds

A disbond indicates an adhesive bond failure between dissimilar materials. This is usually seen as a face sheet disbonding from an underlying sandwich core material. (*See* Figure 10-3.)

In this chapter:
- Types of Damage
- Damage Detection
- Component Identification
- Paint Removal
- Damage
- Damage Removal
- Repair Design
- Vacuum Bagging Materials for Composite Repair
- Repair Instructions
- Scarfing
- The Repair Patch
- Curing Methods

Figure 10-1. Hole and puncture damage to the carbon fiber/aluminum honeycomb nose of an F3 race car monocoque *(left)*; and its appearance after it has been prepared for repair by removing primer and paint *(right)*.

Figure 10-2. Delamination damage.

(Top left and right) This aramid-skinned/foam-cored panel illustrates how the plies of a laminate can actually come apart, or "de-laminate."

(Bottom) Carbon fiber sample shows how a localized impact may cause delamination within the laminate across a significant area.

Figure 10-3. Disbond damage.

(Top) A face skin disbonding from foam core.

(Bottom) Close-up showing that the core was once adhered to the skin.

Core Damage

Core damage can occur with any type of core, and is typically caused by improper vacuum bagging, handling damage in manufacturing, and impacts in-service. Fluid ingress is a common problem with honeycomb core panels. This can facilitate extended damage, from freeze-thaw cycling in aircraft parts.

Resin Damage

Resin damage is caused by many factors, such as fire or excessive heat, UV rays, paint stripper, or impacts. This type of damage may be hard to detect, and it is especially difficult to quantify its effects on the structural integrity of the part. As a general rule, resin damage leads to a greater loss in compressive strength than in tensile strength.

Water Ingression or Intrusion

This is a particular problem with honeycomb cores that causes weight gain, corrosion in aluminum honeycomb, and disbonds if water freezes and expands. It is a very common problem in high-temperature repairs: the heat

Figure 10-4. Core damage to honeycomb core caused by impact.

(Photos courtesy of The University of Auckland/LUSAS)

Figure 10-5. Water ingression damage.

(Left) Note the discoloration of damaged cells in the honeycomb core damaged by water ingression.

(Right) A trailing edge flap from an F/A-18 Navy fighter jet illustrates how moisture-induced disbond damage can be caused during repair. (Photo courtesy of ElectraWatch, Inc.)

 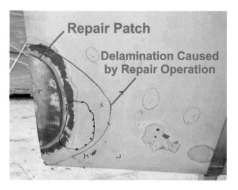

Figure 10-6. Burns cause resin damage, which is easy to see here in the charred circles. However, note the larger areas of delamination caused by the same overheating and much more difficult to detect visually.

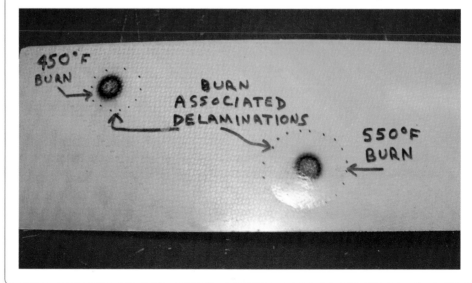

of curing the repair causes trapped water to turn to steam, disbonding face sheets around the repair and ultimately converting a small area of damage into a large one.

Lightning, Fire and Heat Damage

Lightning damage ranges from simple surface abrasions to large through-holes, and often includes thermal damage to resin. Fire or excessive heat may char or burn resin and/or fibers. It may be difficult to determine the extent and depth of this type of damage, but it is extremely important to do so.

Damage Detection

Damage to composites is often hidden to the eye. Where a metal structure will show a dent or "ding" after being damaged, a composite structure may show no visible signs of it and yet may have delaminated plies or other damage within.

Impact energy affects the visibility, as well as the severity, of damage in composite structures. High and medium energy impacts, while severe, are easy to detect. Low energy impacts can easily cause *hidden* damage. (*See* Figure 10-7.)

There are a variety of non-destructive inspection techniques available to help determine the extent and degree of damage. *See* Chapter 7 for illustrations and in-depth descriptions of each.

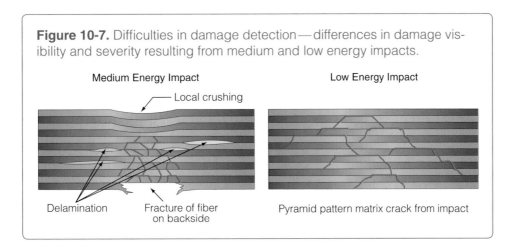

Figure 10-7. Difficulties in damage detection—differences in damage visibility and severity resulting from medium and low energy impacts.

Component Identification

Ideally, composite components should be fully identified before a repair is performed. Such details as material specifications, number of plies and ply orientations, core ribbon direction, ply buildups and drop-offs, and numerous other items need to be clarified and understood before beginning a repair.

For aircraft, this type of information is often available in the structural repair manual (SRM) or equivalent documents. Certain boat manufacturers will supply specific repair instructions and some may provide repair kits. Even if an SRM or equivalent is not available, this type of information is still needed for proper repairs.

Determination can often be done by careful taper sanding through a small sample of the damaged part, and reading the information directly from the composite itself. However, a thorough understanding of composite materials — including weave patterns and ply orientation concepts such as balance, symmetry, nesting, etc. — is mandatory for this type of analysis. (*See* Figures 10-8 and 10-9.)

Repair Materials

A repair should use the same materials, fiber orientation, core orientation, stacking sequence, nesting, curing temperature and sealing as was used in the original fabrication of the part. However, sometimes this is not feasible,

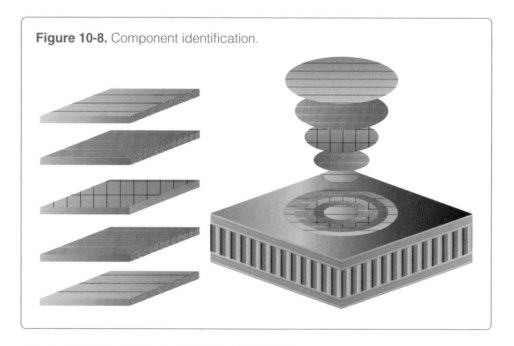

Figure 10-8. Component identification.

Figure 10-9. Component identification by careful taper sanding.

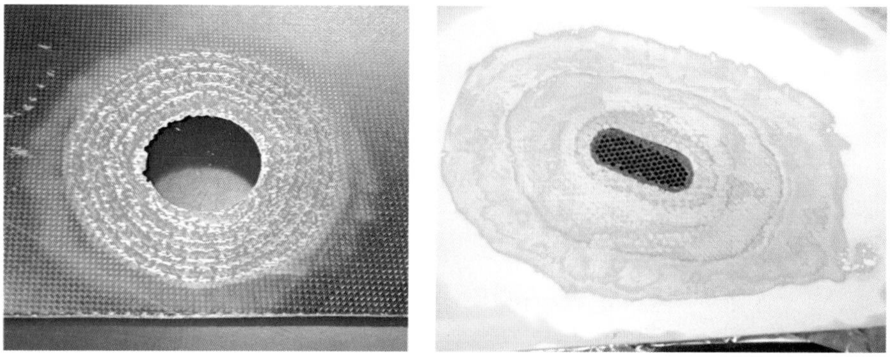

especially in the case that an original structure was manufactured with a fiber/resin system that required an autoclave cure, and the repair is being performed without using an autoclave. Therefore, material substitutions may be allowed or even required by the SRM. In "battle damage" or temporary repairs, often more flexibility in material substitution is allowed than would be normal for permanent repairs.

Nevertheless, the goal remains as much as possible to return the structure to its original strength, stiffness, shape and surface finish, etc. In any composite repair, the following types of repair materials need to be considered and evaluated before the repair is begun, to ensure that the repair is structurally sound:

Matrix Resins

- Resin systems—wet layup vs. prepreg, low-temperature vs. high-temperature.
- Cure cycle requirements and available equipment.

Fibers/Fabrics

- Fiber reinforcement—type of fiber (fiberglass, aramid, carbon, etc.).
- Fiber reinforcement form (unidirectional tape, woven cloth, weave style, etc.).

Core Materials

- Core type—Nomex® honeycomb, aluminum honeycomb, foam, balsa, etc.
- Core orientation—honeycombs have different properties in their ribbon and transverse directions.
- Core adhesive or potting compound.

Lightning Strike Materials

- Mesh and fiber materials—copper or aluminum mesh, conductive fibers, etc.
- Coatings—nickel or ceramic coatings, conductive paints.
- Grounding strips (often bonded to exterior surfaces on aircraft).

Sealants

- Polysulfide (type of coating used on older aircraft parts).
- Polyurethane—newer systems and hybridized coatings that are less brittle, more damage tolerant.
- Rubberized coatings (variety of products to help prevent damage from erosion, etc.).
- Others—boats use gel coats and water-resistant paints; pipes often use thermoplastic coatings for UV and abrasion resistance; sporting goods use a variety of unique finishes.

Paint Removal

The first step is to remove paint and/or outer coatings. Chemical paint strippers must *not* be used, unless you are certain they are specifically designed for composite structures. Most paint strippers are based on methylene chloride, and will attack cured epoxy resin (which is a common matrix resin for composites). Paint and coatings may be removed by:

- Hand sanding.
- Bio-based starch media blasting (wheat, corn, corn hybrid polymers).
- *Careful* grit blasting or plastic media blasting.

It is important to check any repair manuals—such as SRMs for aircraft—or guidelines offered by manufacturers, and also to make sure all health and safety requirements are met.

Chemical Stripping

Most paint strippers are methylene chloride-based. This solvent will attack cured epoxy resin, and is not recommended for composite structures. Alternatives include hand sanding and various particulate blasting methods.

Hand Sanding

Hand sanding is widely used as a paint removal method for composites. No expensive equipment is needed, and the sanding pressure can easily be controlled to avoid laminate damage. Hand sanding is recommended for paint removal in a small area (for example: in an area to be repaired). One must be careful not to damage fibers in the surface ply, unless a specific *sanding ply* has been included on the surface of the component. The obvious disadvantage is the high labor costs, especially on large parts. However, this is by far the most common paint removal method for repair preparation.

Figure 10-10. Hand sanding: Composite parts with paint removed using hand sanding.

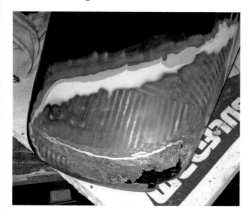
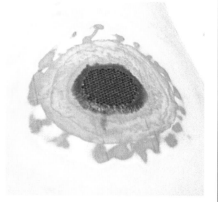

Blasting

There are many types of blasting in use today:

Sandblasting

This type of blasting is possible with composites but must be done very carefully, as it is highly likely that fibers on the top ply will be damaged.

Aluminum Oxide Blasting

Similar to sandblasting, and the same precautions apply.

Plastic Media Blasting

This method is less aggressive than sand or aluminum oxide blasting, and is used often with composites. (*See* Figure 10-11.) While effective, this tech-

Figure 10-11. Blast media.

(Top) Various blast medias for different applications are shown with Plasti-grit® (made from Urea) represented in the middle, as well as the two white piles (made from Clear Cut and Acrylic). The brown-colored pile is walnut shell, the tan colored is corn cob, and the red piles are Nylon and Poly-Carb plastic media. Both plastic media and bio-based starch media *(bottom)* are used to remove coatings from composite structures.

Photo courtesy of Composition Materials Co., Inc.; www.compomat.com

Photo courtesy of Archer Daniels Midland Company

nique has shown some difficulties in the field. The plastic media is most often cleaned and re-used in such systems. This can lead to two problems:

1. The cleaning process is largely designed to remove paint chips and solid matter. If the plastic media becomes contaminated with oil, grease, fuel, etc., these contaminates can be driven back into the composite surface, causing paint adhesion problems.

2. The plastic particles become dull with reuse, and a poorly trained operator may turn up the air pressure to compensate. This can lead to damage of the composite surface, or even blowing a hole through these rather brittle materials.

Starch Media Blasting

This method uses starch media to blast paint from parts. Boeing, the U.S. Air Force, U.S. Army, Cessna, Raytheon, Northrop Grumman, Lockheed Martin, and several other OEMs have approved this method for paint removal from composites. As with plastic media, there can be a contamination prob-

Figure 10-12. Starch media blasting.

(Top) ADM has engineered ENVIROStrip® and eStrip® dry starch-based media to remove multiple tenacious aircraft paint and primer coats from sensitive composite substrates. The dry media process can be used with portable or booth-based blasting systems *(bottom left)*. With some types of primers, ENVIROStrip® can remove coatings down to the final layer, leaving the substrate primer intact *(bottom right)*. (Photos courtesy of Archer Daniels Midland Company)

lem as the starch is reused. This is typically mitigated by limiting the number of reuse cycles to 15 or using dense particle separators.

Starch is much more aggressive when used at higher pressures (over 30 psi). The particles fracture upon impact into smaller particles with increased surface edges. These are even more effective at removing coatings, but can also damage composite substrates. A lower blasting pressure and smaller angle of impingement are recommended for composites, and the media flow range and blasting standoff distance should be carefully selected for the part being treated.

With starch media, there is no chemical attack to worry about, and it is an environmentally benign and biodegradable material that can easily be disposed of. Although very rare and limited to specific areas within the process, dust generated from the wheat starch can possibly explode. This is not possible when standard blast procedures and ventilation are used.

Other Methods

There are many other methods, such as dry ice blasting, Xenon flash lamps, baking soda blasting, etc. Work in this field is continuing, and better methods are currently being researched by many organizations.

Regardless of the blasting system used, one development that is universally accepted is the preference for a flat nozzle with a fan-shaped blast path. This avoids the damaging hot spot in the blast path center of traditional round nozzles. (*See* Figure 10-13.)

Figure 10-13. Flat vs. round nozzle.

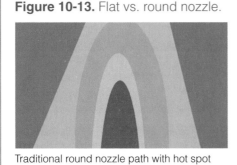

Traditional round nozzle path with hot spot

Preferred fan-shaped blast path of a flat nozzle

Damage Removal

After paint removal, additional damage assessment is performed, because the hidden damage now becomes more apparent. All damaged material must be removed, including anything contaminated by water or other fluids and anything showing signs of corrosion. Be sure to inspect aluminum honeycomb for corrosion. Beginning signs of corrosion include a dull instead of a shiny appearance, and white or gray-colored areas. In advanced corrosion, the honeycomb becomes brittle and tends to flake.

Damage Removal Scenarios

Top Skin Damage Only

Damage is only on the top skin. This situation is usually caused by low-energy impacts and results in matrix resin cracks, fiber breaks, delamination in the top skin, and in a sandwich structure, disbonding of the top face skin from the core.

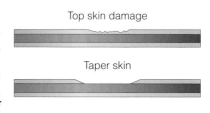

Remove the damaged skin material. Taper sand the surrounding undamaged area to a shallow scarf angle, about $\frac{1}{2}$ inch for each ply in the skin laminate.

Top Skin and Core Damage

Damage extends through the top skin into the core material. This damage results from greater impact and may produce disbonding and/or delamination in the bottom skin, so be sure to perform a thorough inspection. (If the bottom skin has been damaged, then see the final case below.)

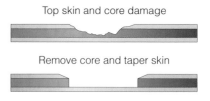

Remove all damaged skin and core material. Surrounding core material must be inspected to ensure it is not contaminated. If it is contaminated, it must be treated or removed. Taper sand the surrounding undamaged area to a shallow scarf angle, about $\frac{1}{2}$ inch for each ply in the skin laminate.

Damage Through the Laminate

In the case of a hole or puncture through the laminate, both top and bottom skins will require damage removal.

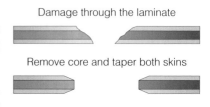

Remove the damaged top skin and core material. Taper sand the surrounding undamaged area to a shallow scarf angle, about $\frac{1}{2}$ inch for each ply in the skin laminate. Remove damaged bottom skin material and taper sand 1 inch for each ply in the laminate.

The tapered areas for the top and bottom skins do not have to match. They should be based on the size of the damage in each skin.

Skin

Damaged composite skin may be removed by careful routing or grinding through the damaged surface. Regardless of the removal method used, always remove damage in circular or oval shapes, and do not make sharp corners. If an irregular shape must be used, then round off each corner to as large a radius as is practical.

Cutting and Peeling

This method was once popular with aluminum honeycomb sandwich parts, but is no longer used due to excessive damage caused to the core. A razor or sharp knife was used to cut through the skin to the honeycomb core. Pliers were then used to peel the skin away from the honeycomb. However, the typical result was significant core damage, as the core would adhere to the skin as it was peeled away, resulting in large areas of ripped and damaged core.

Routing

This is the method of choice for solid laminate damage through the thickness. Diamond-coated router bits give a cleaner cut and reduce splintering

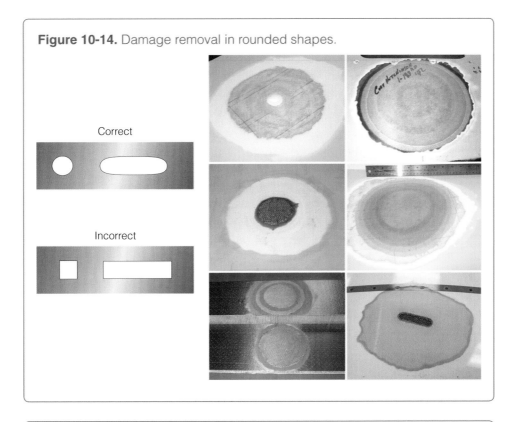

Figure 10-14. Damage removal in rounded shapes.

Correct

Incorrect

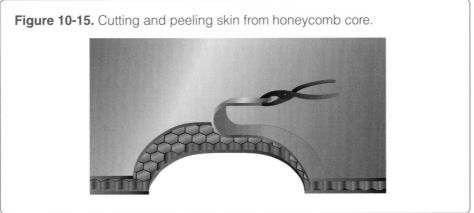

Figure 10-15. Cutting and peeling skin from honeycomb core.

in glass and carbon laminates. For repair of aramid fiber, a tungsten carbide split helix router produces the best results (it minimizes fiber "fuzzing").

Grinding

Also called scarfing or taper sanding, grinding is used for solid laminates or for face sheets on a sandwich laminate. Diamond-coated disks work well for most types of composites. Carefully sand away damaged plies until undamaged plies or the top of the sandwich core is reached.

Core

Cutting and Peeling or Chiseling

This is commonly used with light density core of any type. Use a sharp knife or sharpened chisel/putty knife to make vertical cuts along the edge where the skin has been removed. Be careful not to penetrate or cause damage to the rear skin. Next, slide the knife edge between the core and the rear skin to make a horizontal cut, and then remove the core along its adhesive bondline down to the rear skin. (*See* Figure 10-16.)

Routing and Grinding

Routing may be preferred with cores of higher density. A router may be used to remove the top skin and core; however, be careful not to penetrate the rear skin if it is undamaged. A high-speed grinder is used to carefully remove damaged core until undamaged rear skin is reached. (*See* Figure 10-17.)

Contamination

Damaged composite structures may be contaminated by fuel, hydraulic fluid, oil, etc. Contaminated surfaces must be removed or treated in a way that will enable resins and adhesives to bond.

Solid laminates contaminated with fuel, oil, etc., may be treated by wiping ***thoroughly*** with a solvent, then using reagent grade solvents for the final wipe. It is important to know what the composite is made of, what the contaminant is, and to check any and all suggested procedures to ensure the solvent being used is capable of dissolving the foreign fluid without further damaging the composite structure.

If the core in a sandwich structure is contaminated, replacement of the affected material is the best answer.

Drying

Damaged areas may also absorb moisture, which will prevent achieving a successful composite repair. All affected composite materials must be dried before an effective repair can be achieved. Cured resin as well as fibers will

Figure 10-16. Core removal using a putty knife or chisel.

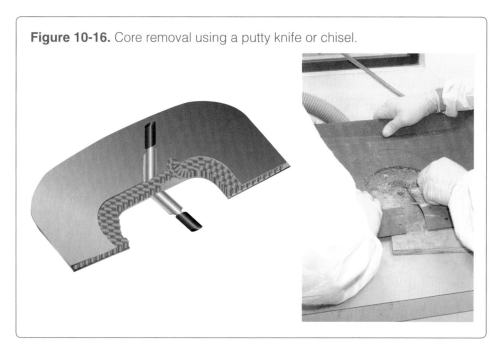

Figure 10-17. Core removal using a high speed grinder.

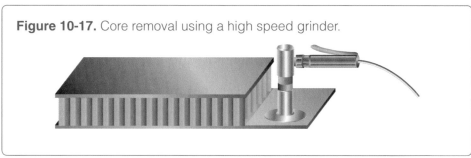

Figure 10-18. Removal of contaminated core.

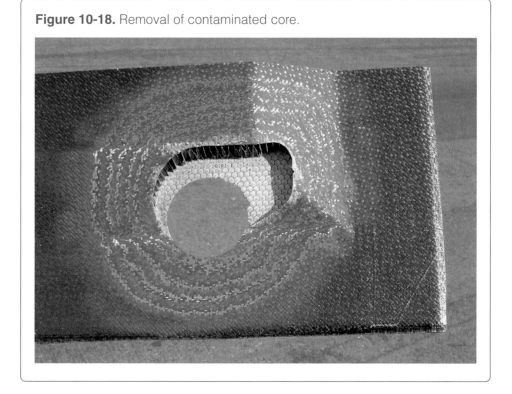

absorb moisture from the environment, and honeycomb cores can hold large quantities of fluid. If performing a repair using high-temperature curing resin or prepreg, all moisture must be removed to prevent steam from forming and disbonding the repair.

When repairing a structure with honeycomb core, the core is almost always moisture contaminated. Refer to the drying cycles specified in repair manuals and guidelines, but note that one hour is usually not enough and sometimes many days of drying time is required. Drying is typically done with a heat blanket and vacuum bag, using heat to convert the moisture to vapor and vacuum to draw it out. Using lower temperatures for longer periods of time achieves the best results. (*See* Figure 10-19 on the next page.)

Guidelines include:

- Drying temperatures vary, but should not be too high: $150-170°F$ is common.
- A rough guideline is to use a drying temperature below the wet T_g (glass transition temperature) of the matrix resin.
- Slower drying using lower temperatures for longer times is best if time permits.
- Bring the part up to temperature slowly and in a controlled process, as rapid temperature rise and high drying temperatures may cause delamination.
- Do not exceed a rate of $5°F$ per minute.

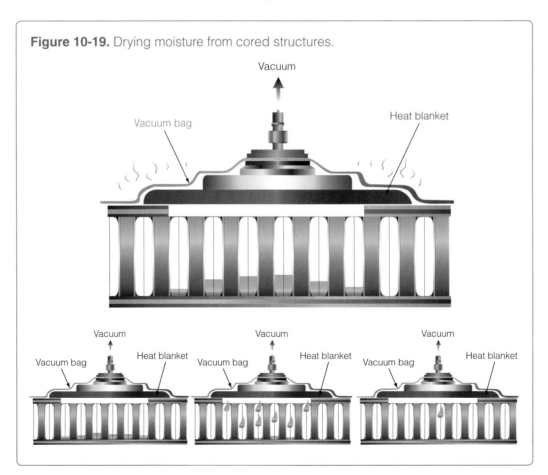

Figure 10-19. Drying moisture from cored structures.

One trick for telling when the structure is dry is to exhaust the vacuum pump through a desiccant that will change color with a change in moisture content. (*See* Figure 10-20.)

A moisture meter (Figure 10-21) may also be used to determine the dryness of composite skins and non-metallic honeycomb core. Moisture meters work best with non-conductive materials; thus, they may not produce good results with carbon fiber composites and metallic cores. Another important consideration is that moisture meters do not give a reliable *calibrated* moisture content reading—i.e., an absolute measurement. However, they can be used for relative readings, such as "is it drier now than it was yesterday?" Note that the readings may vary widely between instruments from different manufacturers.

Figure 10-20. Change in desiccant color as it gains moisture—this blue silica gel turns pink when saturated.

(Photo courtesy of AGM Container Controls)

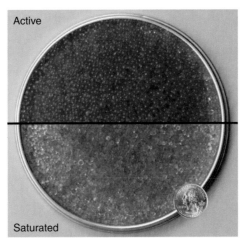

Figure 10-21. This moisture meter from Moisture Register Products works well with cored composite structures.

(Photo at left courtesy of Moisture Register Products)

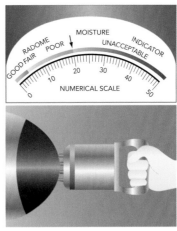

Repair Design

Designing a composite repair is often quite a complex process. This can be seen in Table 10.1, which lists some of the considerations, practicalities and parameters of composite repair.

TABLE 10.1 Composite Repair Design Considerations

Design Considerations	Practicalities	Parameters
▪ Temporary or permanent	▪ Time	▪ Strength
▪ Bonded or bolted	▪ Environment	▪ Stiffness
▪ Flush or external doublers	▪ Equipment	▪ Weight
▪ Single-sided or double-sided	▪ Skills	▪ Shape/Contours
▪ Wet lay-up or prepreg	▪ Materials	▪ Appearance/ Surface finish
▪ Appearance desired	▪ Regulatory and organizational requirements	
▪ Access to part and/ or damaged area	▪ Failure consequences (cosmetic fairing or primary structure)	
▪ Details (shape, heat sinks, etc.)		

The *perfect* repair is to replace the damaged part with a new one. If replacement is not possible, then the ideal repair is to match all original design parameters exactly (e.g. materials, fiber orientation, curing temperature, etc.). In reality, this is rarely possible and compromises are inevitable. However, the design goal of the repair is to return the structure, as much as possible, to its original strength, stiffness, shape and surface finish, etc.

Note: If the extent of damage or overall damaged area exceeds allowable repair limits in applicable repair manuals, then specific engineering support

Figure 10-22.
Repair design considerations.
Is access to both sides needed and available? Repair to aircraft landing gear door *(right)*. What cure temperature is required and what equipment is available for curing? Vacuum-bagged parts being cured with portable bonder *(bottom left)* and heat lamps *(bottom right)*.

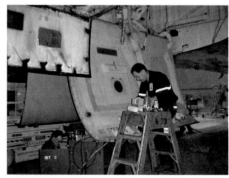

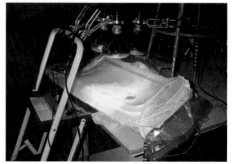

is required in order to proceed with the repair. An individually designed repair is necessary. Since it is such a complex subject, specific training in composite repair design is a *must*.

Structure Types

Repairs to lightly loaded structures might not require as much detailed analysis as those to heavily loaded/primary structures (*see* Table 10-2). However, repair design is still required in order to ensure that the repaired structure will not fail in service.

TABLE 10.2 Composite Structure Types

Structure Type	Description
Primary	Heavily loaded or safety-of-flight structures Example: wing spar
Secondary	Intermediate in load and safety criteria Example: aerodynamic fairing
Tertiary	Lightly loaded or non-critical structures Example: interior sidewall panel

Note that in some cases, a structure is designated as secondary or tertiary from a design load standpoint, but it might be treated as primary structure due to the critical nature and location of the component. An example would be a lightly loaded fairing located in front of the intake of a jet engine. Failure of the fairing itself would not be structurally critical. However, if it gets sucked into the engine, the resulting engine failure could cause more interesting problems.

Types of Repair

Cosmetic

Superficial, non-structural epoxy-based filler is sometimes used to restore a surface. Sometimes this is done with a single ply of glass over the filled area to temporarily keep fluids out and/or hold the filler in. (*See* Figure 10-23.) This type of repair does not regain strength and is used only for small damage areas to secondary or tertiary structures. It does serve to prevent contamination of a core or laminate until a structural repair can be made.

Many non-structural filling compounds are made from polyester resin (autobody filler). They are cheap, readily available and easy-to-use, but they simply are not the right material for composite repair. Therefore even though epoxy-based fillers are more expensive, require more precise mixing, and may need elevated temperature to cure, they are still the material of choice for cosmetic repairs—simply because they stick, and they remain stuck. Epoxy-based fillers also have higher elongation and much lower shrinkage than polyesters, enabling them to resist cracking and perform better long-term.

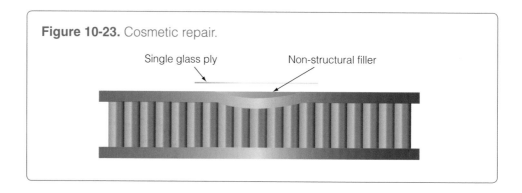

Figure 10-23. Cosmetic repair.

Single glass ply

Non-structural filler

Resin Injection

This type of repair is no longer recommended. The only case where it may be acceptable is for very limited delamination at the edge of a composite part.

For any significant area of delamination or disbond, resin infusion is preferred over resin injection for repair. The lower-than-atmospheric pressure of the vacuum—used in resin infusion—*pulls* resin into cracks and delaminations, re-bonding damage within the laminate. Resin injection, however, uses higher-than-atmospheric pressure to *push* resin into the laminate. Traditionally, resin injection repair techniques specified that vent holes be drilled peripherally around the damaged area in order to prevent further delamination. Even with this measure, the risk is still too great that excessive pressure will force cracks open, grow the delamination, and increase overall damage.

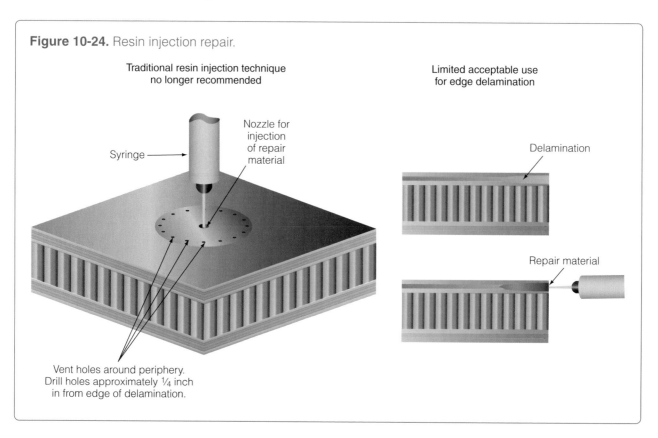

Figure 10-24. Resin injection repair.

Traditional resin injection technique no longer recommended

Limited acceptable use for edge delamination

Syringe

Nozzle for injection of repair material

Delamination

Repair material

Vent holes around periphery. Drill holes approximately ¼ inch in from edge of delamination.

Full structural repairs involving bolted composite doublers can be used in heavily loaded solid laminates. This is often the only practical means of repairing such structures. These patches are often fabricated off of the vessel or aircraft on tooling that controls the contour of the area to be repaired. A liquid shim (paste adhesive) or polysulfide sealant is often required between the patch and the structure to maximize contact and seal against fluid ingress.

Bolted titanium patches are commonly used on thick carbon fiber solid-laminate primary structures because the titanium can easily match the strength and stiffness requirements to transfer loads around the damage. Also, titanium is not susceptible to galvanic corrosion in direct contact with the carbon. Such repairs are not aerodynamically smooth and may cause "signature" problems in structures where low radar cross-section is required; therefore, size and

Figure 10-25. Mechanically fastened composite doubler repair on an external aircraft structure.

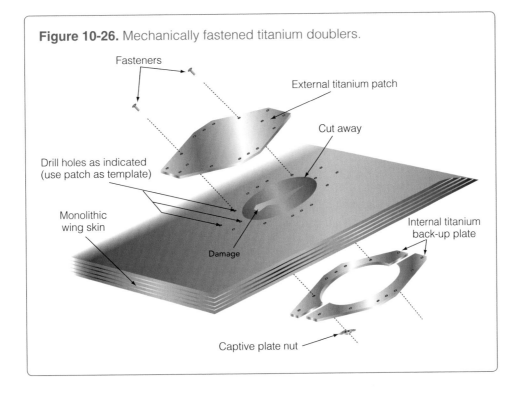

Figure 10-26. Mechanically fastened titanium doublers.

repair of composite structures

215

location limits often apply to these types of repairs. The edges of metallic and composite doublers are always beveled and the corners are rounded to lessen the peak shear-stress concentration around the edges.

This type of repair is quicker and causes less damage to the original structure than a scarfed skin repair. It is also possible to perform while the part is still mounted on the aircraft (with access to only one side of the part). Often these repairs are considered permanent and may require a filler plug in the cleaned-up damage area to satisfy the repair design.

Structural Adhesively-Bonded Doubler Repairs

Bonded external doublers are often used to perform repairs to lightly loaded thin laminate structures, especially with wet lay-up materials. They may require ambient or high-temperature cure cycles, depending on the matrix resin system used for repair. Bonded doubler repairs can regain a significant portion of the original strength of the structure—or even full strength—but with a significant stiffness and weight penalty in many cases. This type of repair is generally easy, relatively quick, and does not require the highly developed skills needed for flush structural repairs.

Flush Structural Repair (Tapered or Scarf Repair)

This type of repair restores full structural properties by forming a joint between the prepared repair area and the repair patch. The repair patch is made by replacing each ply of the composite laminate that has been removed from the damage area. The size of the repair patch should fit exactly the area prepared for repair, except for a final cosmetic or sanding layer, which is often slightly larger to allow for sanding down to achieve a smooth and/or cosmetic surface. Tooling may be required to fabricate the doubler (repair patch) so that it fully conforms to and restores the part's shape and surface.

Flush structural repairs are commonly called *tapered* or *scarfed* repairs, because they require tapering the repair area around the removed damaged plies in preparation for bonding of a co-bonded repair patch. They are common with thin solid laminates and also with sandwich structures, and are currently considered the best type of repair in terms of providing the desired restoration of strength and stiffness with minimal weight increase. They can also be perfectly flush with the original surface if performed carefully. Scarfed repairs are commonly specified by aircraft manufacturers in their SRMs, and recommended as well by boat companies and other original equipment manufacturers (OEMs).

Achieving a good bond between the repair patch and the prepared repair area is critical in completing a successful scarfed repair. One of the basic requirements is careful removal of material in a smooth, flat fashion and at a precise angle, all around the cleaned-up are where the damage has been removed. This scarfing or *taper sanding* is done with a compressed-air-powered high-speed grinder. Practice is the only way to obtain the necessary skill in scarfing required to perform this type of repair.

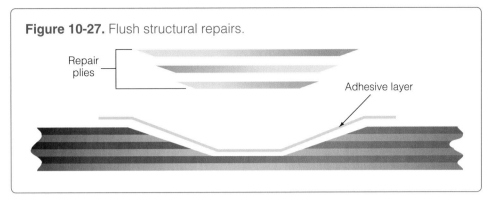

Figure 10-27. Flush structural repairs.

Repair plies

Adhesive layer

Figure 10-28. Scarf repairs.

Scarfing is a vital skill to be perfected for this type of repair, because if done *incorrectly*, it can actually increase the damage to the part.

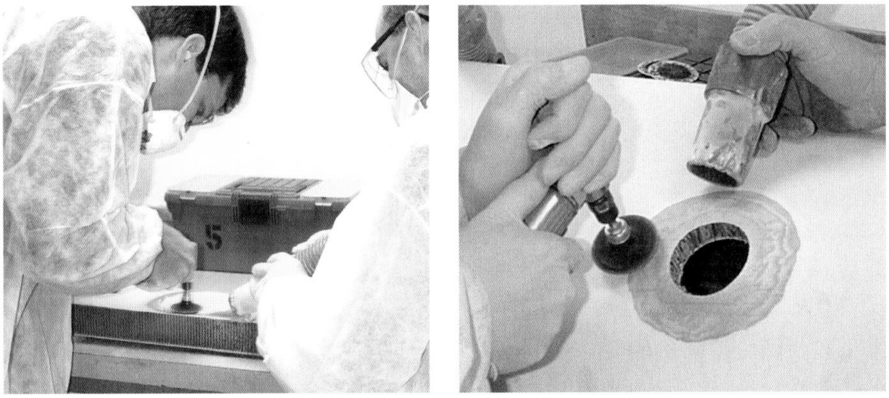

After scarfing, the repair materials (or, "repair patch") are applied to the prepared area, vacuum bagged and cured. Much of the remainder of this chapter will discuss surface preparation, curing and vacuum bagging of scarfed structural repairs.

Resin Infusion Repairs

Note: The information in this section was contributed by André Cocquyt, President of GRPguru.com, and author of *VIP: Vacuum Infusion Process*.

Resin infusion uses vacuum pressure to pull liquid resin into dry reinforcements. Although infusion can be used with 1-sided or 2-sided repairs, it is essential for the mold, tooling or back surface of the repair to hold vacuum. Even the tiniest of leaks will bring air into the repair, potentially destroying its integrity and strength.

Resin infusion is preferred over resin injection for repair. The lower-than-atmospheric pressure of vacuum pulls resin into cracks and delaminations, re-bonding damage within the laminate. Resin injection uses higher-than-atmospheric pressure to push resin into the laminate, which may force open cracks, grow delamination and increase overall damage.

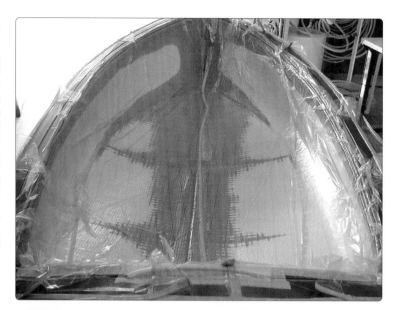

Figure 10-29.
Resin infusion repair.

Boat structures being infused show how flow media enables resin to spread from the infusion ports at a centerline *(top)* and at various locations including hull sides *(bottom)* through the composite laminate indicated by the dark area spreading out *(top)* and up the hull side *(bottom)*.
(Photos courtesy André Cocquyt, GRPguru.com)

The basics of resin infusion repair are the same as with a structural scarfed repair: assess and remove damage, design a repair scheme, and prepare the repair area for bonding. However, in this case repair materials are laid in dry instead of being pre-wet with resin. Part of the repair design strategy is to create pathways for channeling the resin and to make sure it flows completely through the damaged area. The repair area is then sealed and vacuum is applied. Be sure to use epoxy resins formulated for infusion (lower viscosity) and with slow cure times. This enables the resin to completely penetrate the confined spaces of the repair area before it hardens. At the same time, it is important to stay with a system that matches the properties of the original laminate.

Resin infusion is especially useful for repair of large structures (e.g. boat hulls) and with sandwich structures using balsa or foam core. It can significantly reduce the repair time and definitely eliminates the mess of the traditional wet lay-up repair. However, resin infusion repairs require careful planning and significant experience to successfully balance the many variables involved—vacuum, resin pathways, resin cure time, part configuration, etc.

Vacuum Bagging Materials for Composite Repair

How to sequence and apply the variety of materials used in vacuum bagging can be one of the most confusing aspects in performing composite repair. The sequence of materials used is called the vacuum bagging schedule (or simply, bagging schedule).

Vacuum Bagging Requirements for Repairs

To obtain a good vacuum, certain requirements must be met:

- The vacuum bag and the base plate, tooling or mold must **not** be porous—it must be airtight, hold a vacuum, etc.
- Leaks must be eliminated or reduced within a specified range.
- The vacuum pump must be large enough to accommodate the volume of the bagged area.
- A vacuum gauge should be used on the repair to check that the required vacuum is achieved.

The breather is another important factor in achieving good vacuum. The basic rules are:

- There should be a continuous breather layer across the area being vacuum bagged (*see* Figure 10-30).
- The breather should not fill up with resin.
- The breather and bleeder layers (*see* Pages 220–221) should make contact at the edges.

If both the bleeder and breather layers fill up with resin, the dynamics beneath the vacuum bag change. Instead of having atmospheric pressure exerting down-force on the laminate consolidating the plies, the result is a mass of liquid resin under pressure. In this case, the plies are free to float within the resin mass, rather than being consolidated downward toward the tool face. For this reason, a separator layer is often used between the bleeder and breather layers, so that the excess resin being pulled into the bleeder layer does not move up and saturate the breather layer.

Figure 10-30. Bleeder, breather, and bag sequence.

A typical bagging schedule shows the routes for extraction of excess resin (*orange* arrows) and gases (*blue* arrows).

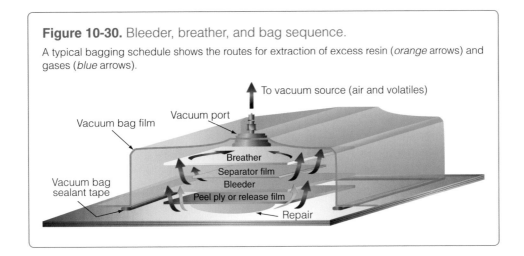

Bag Film and Sealant Tape

The bag film is used as the vacuum membrane that is sealed at the edges to either the surface of the part being repaired, or to the bag film itself if an envelope bag is used. A rubberized sealant tape or putty is used to provide the seal at the periphery. The bag film layer is generally much larger than the area being bagged as extra material is required to form pleats at all of the inside corners and about the periphery of the bag as required to prevent bridging. Bag films are made of Nylon®, Kapton®, Urethane or P.V.A. (poly-vinyl alcohol) materials. They come in a variety of thicknesses, elongations and temperature ratings.

Bleeder Layer

The *bleeder* layer is used to absorb resin from the laminate either through a porous peel ply or a perforated release film as described above. The bleeder layer is usually a 120-style or 1581-style fiberglass fabric. Multiple layers can also be used for heavy resin bleed requirements. This layer usually extends beyond the edge of the lay-up and is secured in place with flash-breaker tape as required.

Breather Layer

The *breather* layer is used to maintain a "breather" path throughout the bag to the vacuum source, so that air and volatiles can escape. It also enables continuous pressure to be applied to the laminate. Typically synthetic fiber materials and/or heavy fiberglass fabric is used for this purpose. The breath-

Figure 10-31. Vacuum bag material and sealant tape.

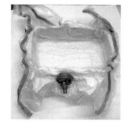

er layer usually extends past the edges of the lay-up so that the edge band makes contact with the bleeder ply around the separator film. The vacuum ports are connected to the breather layer either directly or with strips that run up into the pleats of the bag. It is especially important that adequate breather material be used in the autoclave at pressure.

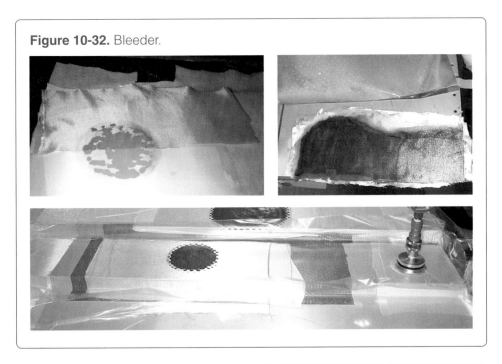

Figure 10-32. Bleeder.

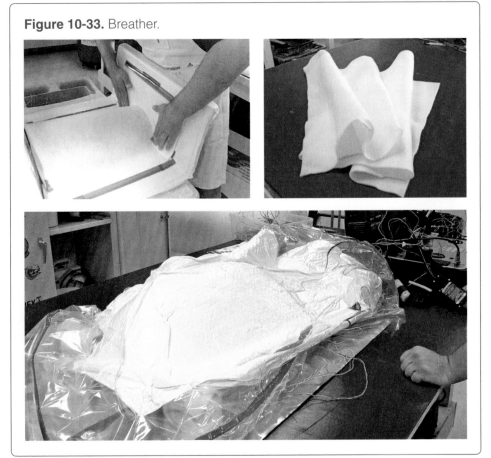

Figure 10-33. Breather.

Peel Ply

Peel ply is a lightweight fabric—typically Nylon or polyester—that is applied on top of the repair patch. After the repair has cured, the peel ply is removed, leaving a smooth surface which is then easy to prepare for painting or bonding. Peel ply is different from perforated film, which may also be used on top of the repair patch. Just because a material "peels" off of the repair does not make it a peel ply. Many peel ply fabrics are treated with a release agent to keep them from sticking to the resin. If the repaired area is to be prepared for subsequent secondary bonding, it is important to choose peel ply fabrics that have *not* been made with release agents.

Perforated Film

This is a release film that has been perforated with a regular hole pattern. It can be used as an alternative to solid release film as a separator film. It may also be used as a filter between two bleeder layers.

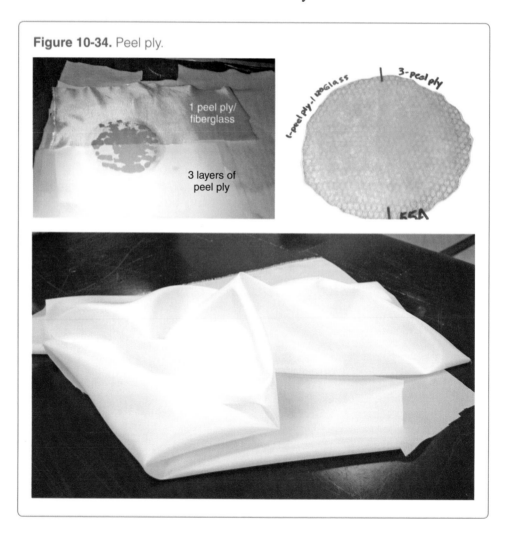

Figure 10-34. Peel ply.

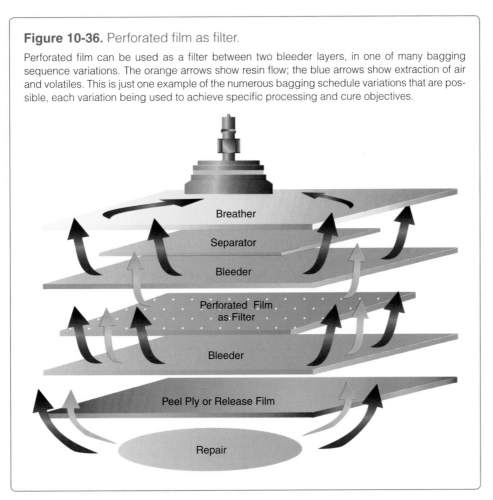

Figure 10-35. Perforated film.

Figure 10-36. Perforated film as filter.

Perforated film can be used as a filter between two bleeder layers, in one of many bagging sequence variations. The orange arrows show resin flow; the blue arrows show extraction of air and volatiles. This is just one example of the numerous bagging schedule variations that are possible, each variation being used to achieve specific processing and cure objectives.

Breather

Separator

Bleeder

Perforated Film as Filter

Bleeder

Peel Ply or Release Film

Repair

Separator Film Layer

The separator layer is used between the bleeder layer and the subsequent breather cloth to restrict or prevent resin flow. This is usually a solid or perforated release film that extends to the edge of the lay-up, but stops slightly inside the edge of the bleeder layer, to allow a gas path to the vacuum ports. Non-porous FEP can also be used as a separator layer.

Vacuum Gauges

Vacuum gauges show the actual pressure achieved at the repair. They should be located on top of the repair, but on the opposite side from the vacuum port. Vacuum gauges are typically removed during high-temperature cures, as they can be damaged by the high temperatures.

Vacuum Ports

Vacuum ports are used to connect the vacuum bag to the vacuum source. Quick-disconnect fittings attach to the vacuum ports, enabling vacuum hoses to be easily connected and disconnected without losing vacuum in the bag. The ports are installed as follows:

1. Place the base plate(s) on top of the breather cloth at desired location(s). Cover with bagging film, cutting a small slit through the film directly over each base plate.

2. Lock the top plate into each base plate through the slits in the bagging film. The gasket on the underside of the top plate enables an air-tight through-hole in the bagging film.

3. Connect the quick-disconnect fittings to the top-plates.

4. Connect the vacuum hose into the quick-disconnect fittings.

Large repair areas may require more than one vacuum pump with several extraction points. A general rule of thumb is to use one vacuum port for each square yard of surface being vacuum bagged.

It is also recommended to run a leak test by pulling at least 20 in. Hg prior to curing. Watch the vacuum gauge for 2 to 5 minutes, evaluate any loss in vacuum and minimize it. Ultrasonic leak detectors are a relatively cheap way to

Figure 10-37. Separator film layer.

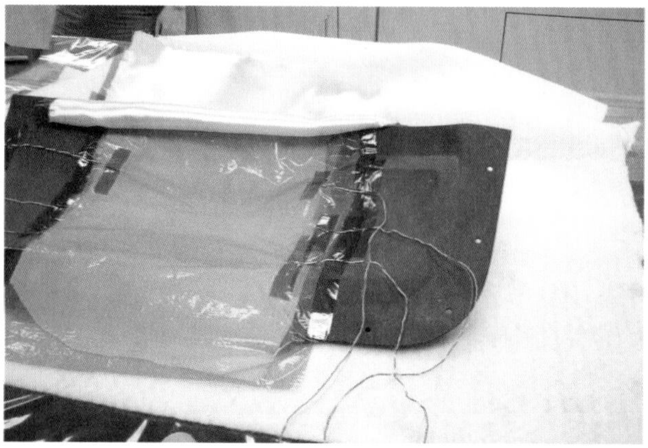

Figure 10-38. Vacuum gauges should be located on the opposite side from the vacuum port *(right)*.

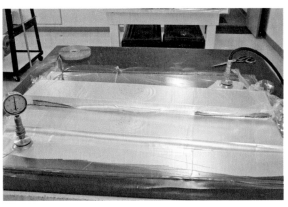

Figure 10-39. Vacuum ports.

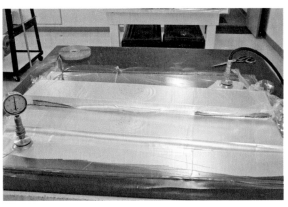

double-check vacuum integrity—a few hundred dollars can save thousands in a failed repair. (*See* Figure 10-40.) This time-saving device amplifies the ultrasonic frequencies of leaks while filtering out background noise. Sweep the perimeter of the bagged area with the device probe, listen for the rushing sound of leaks through the device headphones, and check the LED display which indicates leak strength. Minimizing leaks is very important and can be accomplished with persistence.

In addition to bagging schedules, there are other issues in vacuum bagging:

- Thermocouple quantity and placement
- Caul plates
- Heat blanket issues
- Heat sinks
- Extra adhesive layers
- Vacuum port quantity and placement
- Bagging "pleats"
- How much vacuum to draw

There are no quick and easy answers here. The bagging schedules used with wet layup repairs are affected by the ambient temperature as well as the pot life and amount of resin used. With prepreg repairs, the shelf life of the prepreg as well as its out-time "B-stage" affect the bagging schedule. Also, older but still certified prepreg may require a different schedule than fresh prepreg. The best practice is to test first whenever in doubt.

Figure 10-40. Checking for vacuum leaks—ultrasonic leak detectors help to pinpoint vacuum bag leaks.

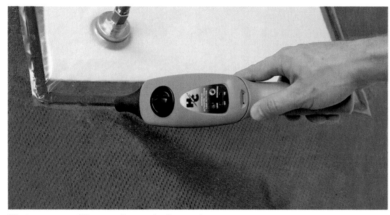

(Photo courtesy of Heatcon Composite Systems)

(Photo courtesy of Fibre Glast Developments Corp.)

Repair Instructions

Below is a review of the general steps in performing a composite scarfed repair with a flush, bonded repair patch.

1. Get best access possible, both sides if feasible.

2. Inspect for extent of damage:

   - Visual
   - Tap, ultrasonic, X-rays, etc. *See* Figure 10-41.

3. Remove all damaged and contaminated material. *See* Figure 10-42.

   - Remove damage in a circular, oval or rounded pattern.
   - Remove or treat contaminated material.

4. Determine the part's ply count, orientations, laminate thickness and materials in preparation for repair design and scarfing. There are several choices depending on what information and materials are available.

   - Thorough repair instructions will detail exactly the number of plies, ply orientations and material specifications for the repair, as well as the scarf diameter for the repair area.
   - Ply count, ply orientations and material specifications may also be found in documentation of the original structure's manufacture (engineering drawings).
   - To determine plies, orientations and material practically, there are 2 options:

     —If a large enough piece of the damaged material is available, scarfing into that will reveal the number of plies, etc.

     —Otherwise, it will be necessary to taper sand into the area around the damage, roughly 1/2 inch for each ply of laminate thickness. Use calipers to determine the laminate thickness.

5. Taper sand/scarf the repair area, according to repair design instructions, to create a smooth, flat surface with high surface energy. *See* Figure 10-43.

6. Thoroughly dry the structure if moisture is present.

7. Develop a repair design, based on the damage and original structure information. Engineering support is usually required to ensure a successful repair.

8. Replace materials. *See* Figure 10-44.

   a. Solid laminate

   - Adhesive layer first.
   - Repair plies—match orientations with original structure.
   - Extra plies—usually orientation matches original outer ply.
   - Outer sanding layer if required (usually a film adhesive or one-ply fiberglass fabric layer).

b. Through-damaged sandwich structure

- Adhesive layer.
- One or more filler plies.
- Core material.
- Core splice adhesive around core plug.
- Bottom skin plies.
- Top skin plies.
- Outer sanding layer if required.

Note: A two-step process is recommended for sandwich panel structures where the core is bonded then machined flush, followed by the skin repairs. The top and bottom skin repairs follow solid laminate guidelines above.

9. Vacuum-bag and cure repair plies as required. *See* Figure 10-45.

10. Inspect repair. *See* Figure 10-46.

11. Sand and finish as required. Do not sand into fibers of repair plies.

Figure 10-41. Ultrasonic inspection.

A-scan *(top)* and C-scan *(bottom)* ultrasonic scans help to define the extent and degree of damage.

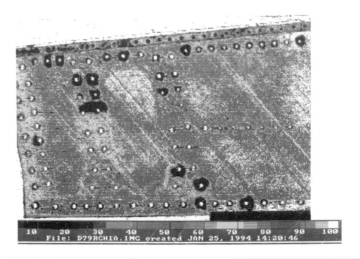

Figure 10-42. Remove damaged material.

Damaged core is cut and scraped away. It may also be removed with an air-powered grinder.

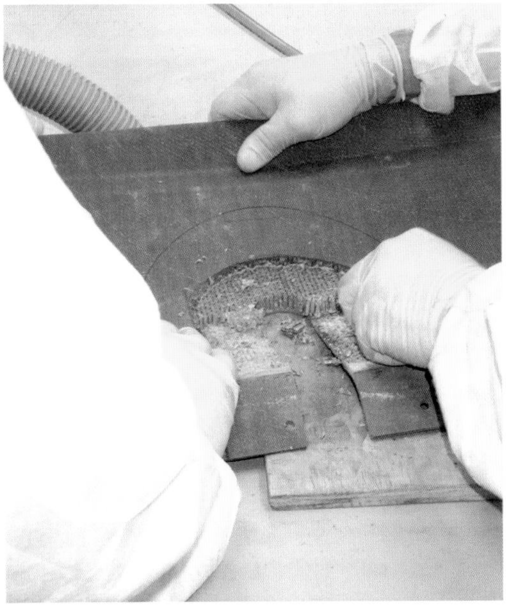

Figure 10-43. Taper sanding/scarfing.

Figure 10-44. Replace materials.

First, bottom skin plies are replaced *(top)*, and then core material *(bottom)*.

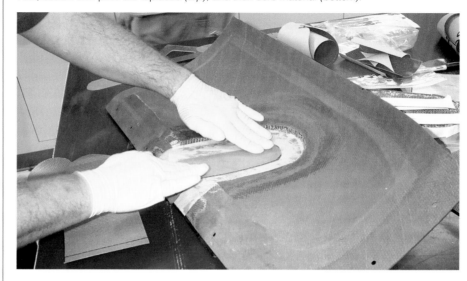

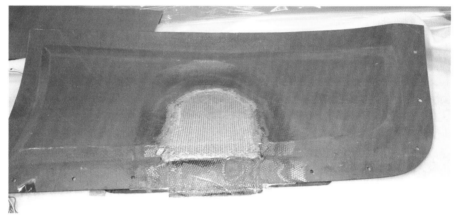

Figure 10-45. Vacuum bag and cure repair.

Vacuum bagging materials are sequenced on top of the repair *(left)* and then sealed within the vacuum bag *(right)*. The repair is then cured using heat and vacuum pressure.

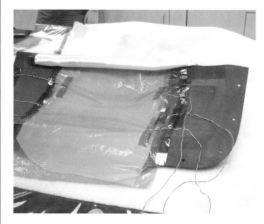

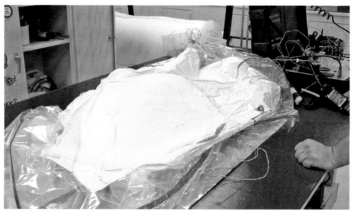

Figure 10-46. Inspect repair.

Inner skin *(top)* and outer skin *(bottom)* are inspected. Note use of repair tooling in restoring curvature of bottom skin surface.

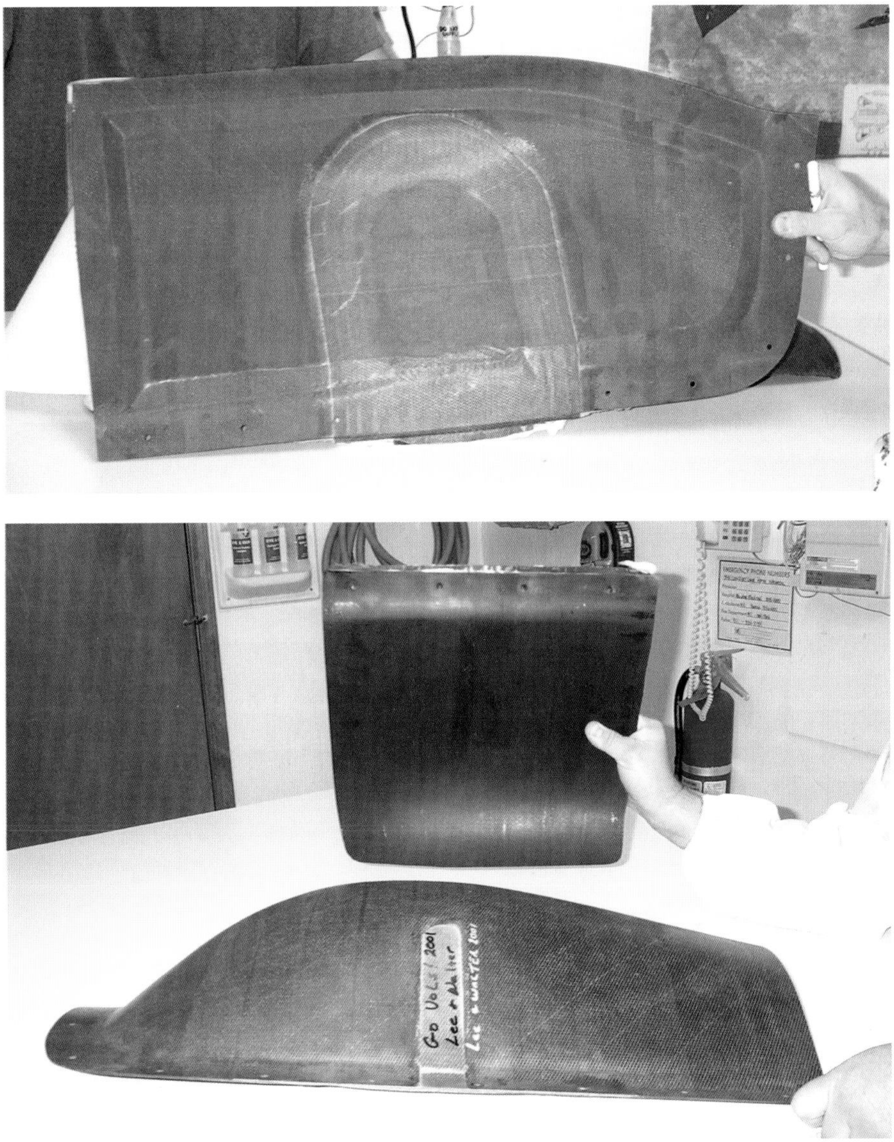

Scarfing

After completing initial damage removal, the area around the repair must be prepared. The corners of the damage removal hole must be rounded off and the area beyond this should be tapered to provide the best load transfer when the repair patch is bonded in. Scarfing, or taper sanding, is usually achieved using a compressed-air-powered high-speed grinder. This is a gentle process, which prepares the damaged area for application of a repair patch. It is imperative to follow all repair manual guidelines, and significant skill and practice on the part of the repair technician is mandatory.

Note: If the damaged area exceeds the allowable repair limits in the governing repair manuals, then specific engineering support is required in order to proceed with the repair.

Tapered (scarf) repairs are typically circular or oval. There are two ways to specify how much area should be tapered around the damage: (1) tapered-scarf distance per ply, or (2) tapered-scarf angle.

Tapered (Scarf) Distance per Ply

A rough rule-of-thumb for the amount of material to remove during scarfing is to taper sand approximately $\frac{1}{2}$-inch of area per ply of composite laminate. In the example below, following this rule would result in scarfing $\frac{1}{2}$-inch for each of 4 plies to produce a scarf distance of 2 inches. Scarf distance is measured from the inner edge of the scarf to the outer edge of the scarf. The inner edge of the scarf begins where the damage has been removed. Counting the hole diameter of 0.5 inch, the scarf diameter would be 4.5 inches. The scarf diameter is the diameter of the whole scarfed area. Typical scarf distances range from 12 to 120 times the thickness of the removal.

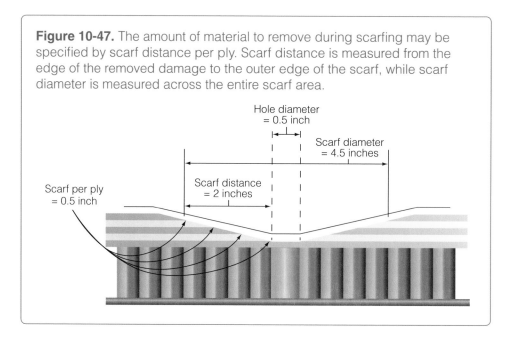

Figure 10-47. The amount of material to remove during scarfing may be specified by scarf distance per ply. Scarf distance is measured from the edge of the removed damage to the outer edge of the scarf, while scarf diameter is measured across the entire scarf area.

Tapered (Scarf) Angle

The tapered or scarf area may also be specified by a scarf angle, rather than distance per ply. Scarf angle is defined as the scarf distance divided by the thickness of the laminate, and is always given as a ratio to 1 (e.g. 12:1, 20:1, etc.). In the Figure 10-48 example, a 0.25-inch thick laminate is to be tapered with a 20:1 scarf angle ratio. To calculate the scarf distance required, simply multiply the laminate thickness by the scarf angle (0.25 inch x 20 = 5 inches).

The steeper the scarf, the less undamaged material is removed. Lightly loaded structures may be able to tolerate a smaller, steeper scarf. A typical scarf

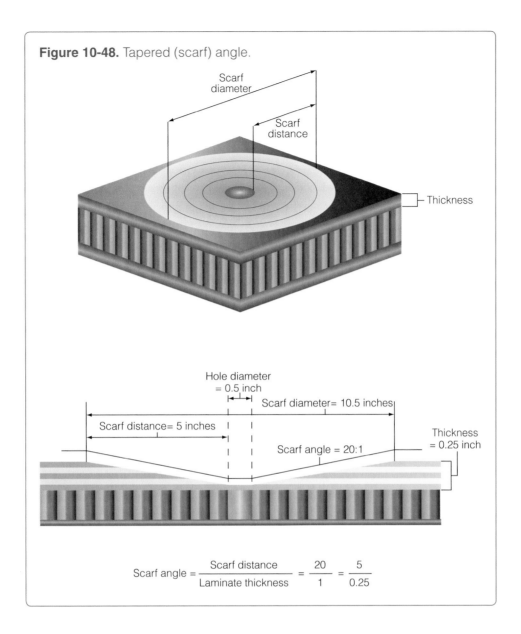

Figure 10-48. Tapered (scarf) angle.

$$\text{Scarf angle} = \frac{\text{Scarf distance}}{\text{Laminate thickness}} = \frac{20}{1} = \frac{5}{0.25}$$

angle for lightly loaded non-aerospace structures is 12:1 (~5°). The flatter the scarf (more area per ply), the larger the adhesive bond is and the lower the load per square inch on the bond. Heavily-loaded structures usually require a larger, gentler scarf. For example, aerospace structures require a scarf angle of at least 20:1 (~3°), and thin, heavily-loaded primary structures may require a scarf angle of 50:1 (~1.3°).

Scarfing vs. Stepping

Stepping is an alternate method for removing material in preparation for applying a repair patch. In stepping, the overall angle is achieved by removing a precise area of material per ply of composite laminate. The result is a "terraced" slope as opposed to the smooth slope of a tapered (scarf) area. Stepping is more difficult to perform and less forgiving. It also risks extending delamination and causing stress concentrations at the edges of the steps.

Most consider the tapered (scarf) method to be easier, and it is generally considered to be better than step preparation because when preparing a stepped repair, it is very difficult to cut cleanly through a ply without damaging fibers in the ply below. Also, great care must be taken not to extend delamination as each ply is dug out or peeled away. The sharp edges of the steps can also lead to abrupt stress concentrations in the repaired laminate, increasing the possibility of failure. In general, a stepped repair is more difficult to perform and less forgiving with regards to precision and quality. This has led to reduced use of this technique and a preference for tapered or scarf repairs.

Figure 10-49. Scarfing vs. stepping.

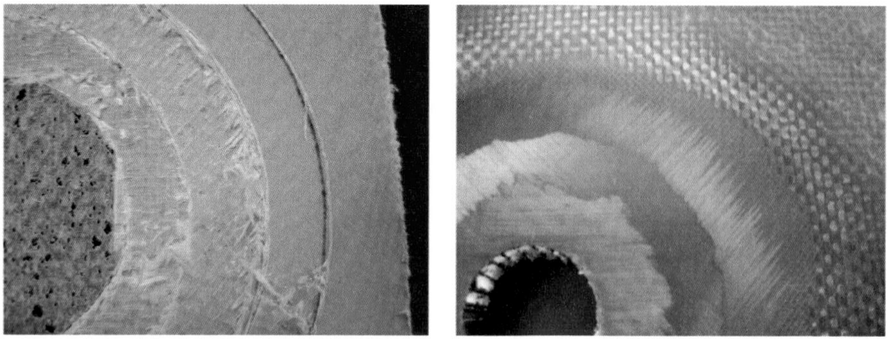

Step-removal on Kevlar skin

Tapered (scarf) on carbon skin

The Repair Patch

The plies of the repair patch are cut to fit the prepared repair area and should have rounded corners. The repair patch is attempting to replace the damaged area in the composite laminate exactly, restoring it as much as possible to original. (*See* Figure 10-50.) Thus, the number of plies and orientations of each ply must match, layer for layer, that of the original structure. Each ply in the repair patch matches the corresponding layer in the scarfed repair area.

Therefore, in an exact ply-by-ply replacement, will the repaired structure be as strong as the original? No. Depending on many details, the repaired structure is typically about 70-80% as strong as the original undamaged structure.

Is it possible to make a repaired structure as strong as the original? Yes. However, extra repair plies must be added to compensate for the loss of strength caused by the repair. This means the repair will not be perfectly flush, and also that the repaired structure will be stiffer than the original.

Is the extra stiffness a problem? If the original structure is not stiffness-critical and is primarily loaded in straight tension or compression, then a stiffer repair will most likely be fine. However, if the structure flexes significantly under load, the stiffened area of the full-strength repair may contribute to failure of the repair. Some repairs may therefore need to be deliberately under-strength, in order to match the stiffness of that original structure.

Figure 10-50. The repair patch must match the material of the original structure, layer for layer.

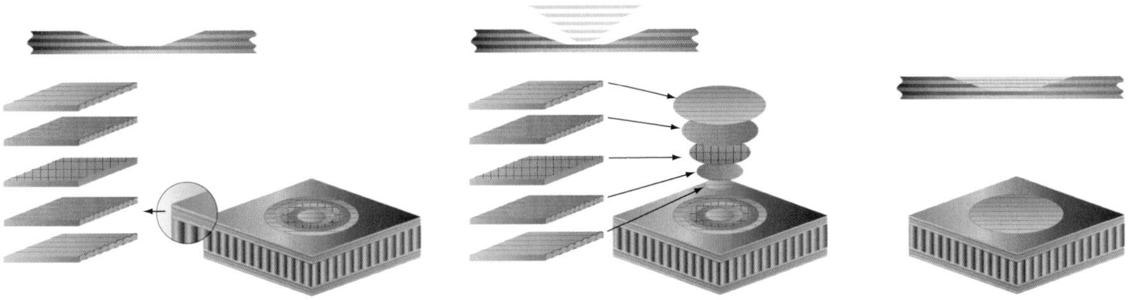

Figure 10-51. Extra stiffness—when the repaired structure flexes significantly under load, a repair that is stiffer than the original structure may fail at the edges.

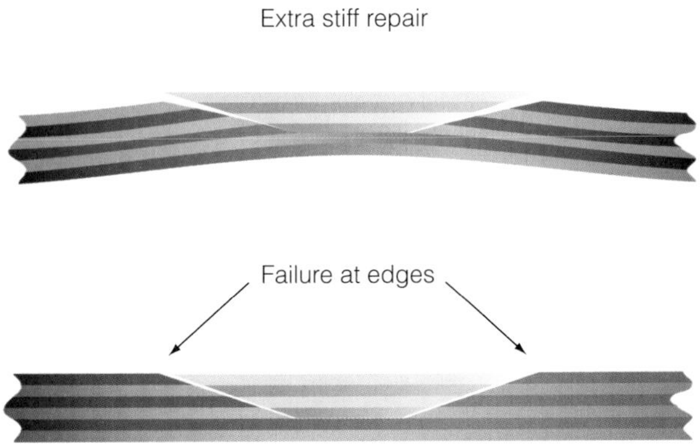

Extra stiff repair

Failure at edges

Curing Methods

Hot Bonders

Portable repair equipment is commonly referred to as hot bonders. Produced by a variety of manufacturers, most hot bonders are suitcase-sized or smaller and are used to control the application of heat and vacuum to a composite repair. They are especially useful for field repairs, in situations where it is not possible to remove the damaged part for repair. They also can be used to monitor and control temperature in oven-cured repairs. (*See* Figure 10-52.)

Hot bonders are most commonly used to control heat blankets, but can also be used to control heat lamps, hot air guns, or even ovens. They can be programmed to store as many as 30 different cure cycles in memory. They require an electrical power source, and control the heating through thermocouples.

Hot Bonders Rely on Thermocouples

A thermocouple consists of two wires of different metals that are welded together at one end and joined at the other end by a plug. The wires at the

repair of
composite
structures
235

Figure 10-52. Hot bonders.

Brisk Heat ACR3 dual zone hot bonder *(top left and middle)*; WichiTech HB-1 single zone *(top right)* and WichiTech HB 2 dual zone *(bottom left)* hot bonders; Heatcon Composite Systems dual-zone hot bonder *(bottom right)*.

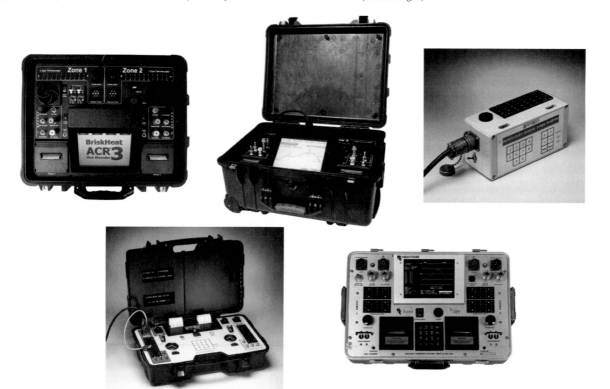

(Photos courtesy of BriskHeat, WichiTech, and Heatcon Composite Systems)

Figure 10-53. Hot bonders control heat lamps and blankets.

A composite repair using a hot bonder and heat lamps *(top)*. Heat blankets come in a variety of different shapes and sizes *(bottom)*.

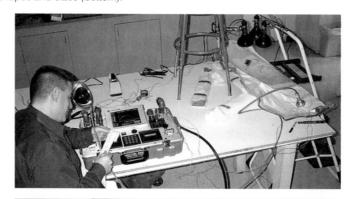

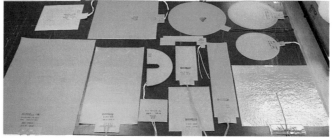

plug end are held at a fixed lower temperature. When the wires at the welded end are heated, a current is generated, which varies according to the temperature difference between the welded (hot) and the plug (cold) ends of the circuit. A voltmeter in the hot bonder translates this current into a very accurate temperature reading — up to 700°F for the "J"-type thermocouples most commonly used with hot bonders. (*See* Figure 10-54 on the next page.)

Thermocouples are placed around the repair area to monitor the temperature of the repair. Based on those temperature readings, the hot bonder's controller processor determines whether to turn the heat source (e.g., heat blanket, oven, heat lamps, etc.) ON or OFF in order to adjust the temperature at the thermocouples, and maintain the programmed cure cycle parameters.

Cure Cycle

The cure parameters form the cure cycle, which is pre-programmed into the hot bonder. Most hot bonders can store multiple cure cycles in memory. Operators select the cure cycle desired by entering a number on the keypad, similar to speed-dial. Cure cycles tell the hot bonder how fast to warm up (ramp), how long to hold the cure temperature (soak), and how fast to cool down. Cure cycles will also specify vacuum pressure. Cure cycles for hot bonder repairs typically have a single soak as opposed to those for autoclave repairs which often have two soaks. (*See* Figure 10-56 on Page 239.)

A multiple-zone hot bonder (usually two zones) can run more than one cure at a time, or control a single repair cure over a larger than normal area. What constitutes a "large area repair" can range from 3' x 3' to 2' x 8', and is typically limited by the size of heat source available (e.g., heat blanket or heat lamps). 120-volt heat blankets are limited to about 5 square feet on a 30-amp circuit. 240-volt blankets can cover about a 10-square-foot area.

In addition to a power source, hot bonders also require a clean compressed air source. This is used in conjunction with a built-in venturi to generate vacuum for compacting the repair plies during cure to achieve a strong adhesive bond.

Hot bonders are useful in controlling the curing process for a composite repair, but they are not cheap. Costs typically range from $5,000 to $30,000. Also, by definition, using a hot bonder means an elevated cure temperature. This requires well-trained personnel who are familiar with the many issues involved in elevated-temperature composite repairs.

Hot Bonder Features

Keypad and display: The keypad enables the operator to input commands and program a variety of different features, including cure cycle parameters, alarms, and repair zones. The display guides the operator in controlling the curing process. Many units will constantly display the cure status in both graphic and text form, and will also provide text descriptions for alarms.

Chart recorder: The chart recorder prints a log record of each repair cure cycle, providing evidence that the correct temperature and vacuum was used

Figure 10-54. Hot bonders and thermocouples.

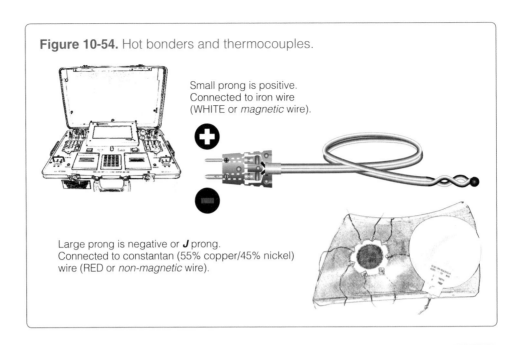

Small prong is positive.
Connected to iron wire
(WHITE or *magnetic* wire).

Large prong is negative or ***J*** prong.
Connected to constantan (55% copper/45% nickel)
wire (RED or *non-magnetic* wire).

Figure 10-55. Hot bonder temperature controller.

This hot bonder has been programmed to maintain a 200°F cure. It will turn each heat lamp ON if the average temperature for that zone's thermocouples reads *less than* 200°F. It will turn the heat lamp OFF if that zone's thermocouples return an average that is *more than* 200°F.

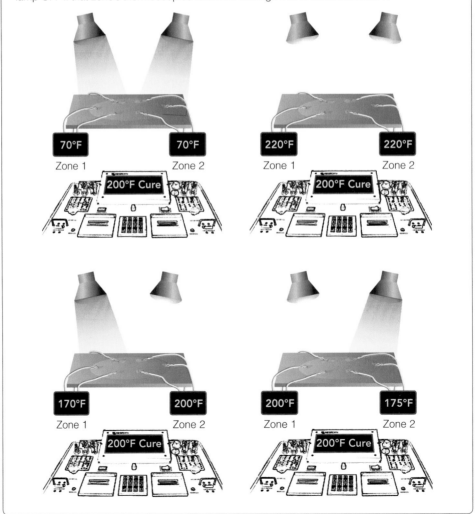

Figure 10-56. Typical hot bonder repair cure cycle.

This hot bonder repair cure cycle features a ramp to a 350°F soak and an hour-long cool-down.

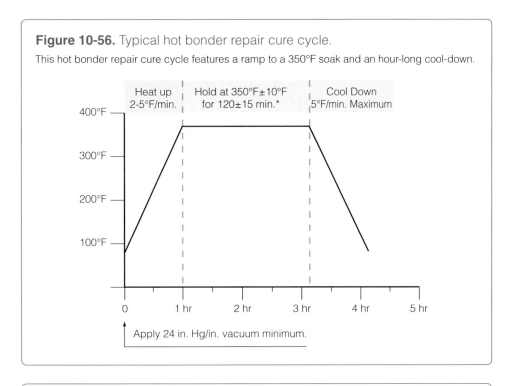

Heat up 2-5°F/min. | Hold at 350°F±10°F for 120±15 min.* | Cool Down 5°F/min. Maximum

Apply 24 in. Hg/in. vacuum minimum.

Figure 10-57. Dual-zone hot bonders.

Repair of two composite parts using a two-zone hot bonder *(left)*, which prints a log record for each zone *(right)*, i.e. a separate log record for each of the two parts being repaired.

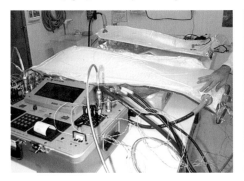

Figure 10-58. Hot bonder features.

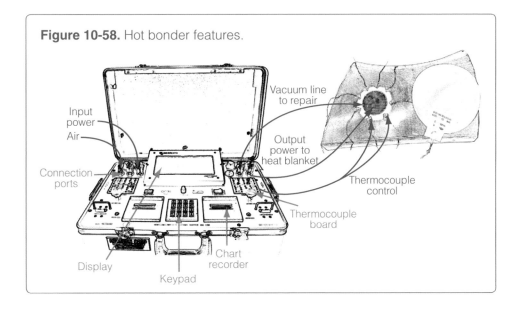

Input power

Air

Connection ports

Display

Keypad

Chart recorder

Vacuum line to repair

Output power to heat blanket

Thermocouple control

Thermocouple board

repair of composite structures

239

throughout the repair. This is required for FAA documentation. Print-out data can include time, date, cure cycle description, tag number, part number, operator ID, actual temperature and vacuum at selected intervals, graph of cure showing duration and temperature, description of alarms and alarm clear, and in-process program changes.

Alarms: Alarms alert the operator when cure cycle parameters are not being met or when conditions arise which may threaten the repair. These include low temperature, low vacuum, over-temperature, thermocouple failure, thermocouple fall (declining temperature), heat blanket failure, and power interrupt.

Vacuum Pump: Most bonders contain an internal venturi that transforms compressed air into a vacuum of 25–28" Hg. Some hot bonder models use an internal electric pump instead of a venturi.

Connection Ports: Connection ports on a hot bonder typically include input power (power supply), output power (to heat source), air intake (compressed air or plant air), vacuum line (to repair), and thermocouple control board.

Figure 10-59. A hot bonder display showing a graphic of the cure cycle being used.

Heat Blankets and Other Heating Methods

Accurate temperature control and uniformity are difficult with heat lamps and hot air guns. Temporary ovens and heat blankets offer better control of curing temperature and even distribution of heating.

A heat blanket the same size as the repair area is too small. Heat blankets must be considerably larger than the area being cured. For example, an 8-inch diameter circular repair will require at least a 12-inch diameter heat blanket. Using this size of blanket allows for thermocouple placement well inside the blanket edges.

Controlling thermocouples placed near the edge will make the blanket overheat, causing a very real fire risk and a damage risk to the component. Also, temperatures drop within 2 inches of the blanket's edge. At the very edge of the blanket, the temperature is easily 100°F colder than at its center. This is often referred to as the heat blanket's "cold zone."

Figure 10-60. Repair curing methods.

Composite repairs may be cured using heat lamps *(top left)* and hot air guns or blowers *(top right)*; however, heat blankets provide the best control and distribution of curing temperature *(bottom)*.

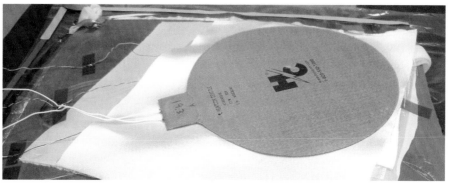

Figure 10-61. Thermocouple placement.

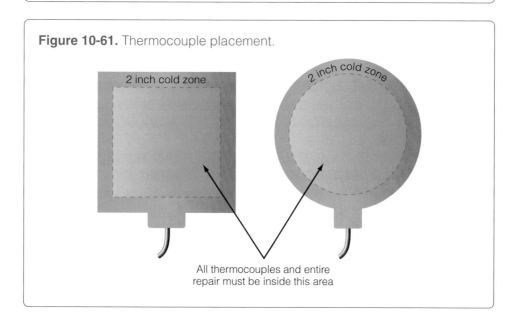

2 inch cold zone

2 inch cold zone

All thermocouples and entire repair must be inside this area

Cure Temperature Considerations

There are four common choices, with variations, as shown in Table 10.3.

TABLE 10.3. Common Cure Temperatures

Curing Temperature	Manufacturing Method
Room temp – 77°F (25°C)	Wet lay-up
Room temp with post-cure at 150°F (66°C) – 200° (93°C)	Wet lay-up
250°F (121°C)	Prepregs
350°F (177°C)	Prepregs

Which is best? It depends. Post-cures are very often required for room-temperature laminating resins, in order to develop full strength in a reasonable time. Often, but not always, prepregs cured at higher temperatures are stronger than room temperature cured materials. However, high curing temperatures can cause problems:

- Steam creation in laminates and over cores
- Blown skins
- Excessive porosity in bondlines
- Uneven heating problems
- Overheating/fire risks
- Increased documentation, often with expensive equipment
- More training required.

Often, repair manuals will offer a choice, especially with smaller repairs. In this situation, go with the lowest-temperature repair allowed by the manual or guidelines. It will offer an easier repair with less ways for things to go wrong. If there is no choice, then it is imperative to do exactly what is prescribed in the manual or guidelines.

Thermocouple Placement and Other Issues

Proper thermocouple placement is crucial for proper high-temperature repairs using hot bonders. When heat blankets are used, multiple thermocouples are required. The general guideline is: the more thermocouples, the better. This is because thermocouple failures are common and replacement during a running cure is virtually impossible.

Another reason for using numerous thermocouples is to control temperature spread across the repair. Remember, the goal is to supply uniform heat to the repair and avoid high spikes and/or cold spots. Temperatures can vary widely across a repair during cure. Spreads of 60–80°F are not uncommon and can be caused by the factors below.

Variations in Part Thickness and Conductivity

Most composite parts contain variances in thickness and/or temperature conductivity of the materials. Thicker laminates typically absorb heat while thinner surface skins will shed it rapidly. This can result in a wide temperature variation during cure.

Heat Sinks

There may also be heat sinks in the part or underlying tooling. For example, thick spars in a structure will take longer to absorb heat on the ramp up and will also retain heat longer on the cool down.

Non-uniformity in Heat Blankets

Heat blankets are made by embedding wires between two layers of silicone rubber. As heat blankets are conformed to complex shaped parts, wires may be stretched around a corner. This area of stretched, thinner wire will create a hot spot in the heat blanket.

Thermocouple Placement—4 Rules

Here are four basic rules for installing thermocouples around a repair:

1. A thin aluminum (or composite) caul plate will help maintain uniform temperature over the repair. Place the caul plate over the repair/bleeder stack and below the heat blanket to equally distribute the heat from the heat blanket to the repair and the thermocouples. Thermocouples should be taped to the top or back surface (surface away from the repair) of the caul plate. *Never place thermocouples under a caul plate.*

2. Always provide a heat blanket that is a minimum of 4 inches larger than the maximum repair size. Locate the thermocouples well inside the heat blanket. This will prevent errors in temperature readings and problems that can result from being in the 2-inch cool zone at the blanket's edge. However, do not place thermocouples directly over the repair area (when **not** using a caul plate) as they will leave an imprint or "mark-off" once vacuum is pulled.

3. Always place a layer of flash tape under the head of each thermocouple around the repair to insulate the thermocouple from the conductive (carbon or metal) surface. This will prevent thermocouple cross-talk and small spikes in the thermocouple readings. Also, tape directly over the head with one layer of flash tape to secure each thermocouple in place.

4. Always use a uniform amount of bleeder/breather materials under the heat blanket, and insulation above the heat blanket. Never double-up the insulation just to fit it inside the bag. This could provide more insulation over one thermocouple and less over another. Following these basic tips will reduce spreads in thermocouple temperature readings, help to prevent erroneous readings, alarms and runaways, and result in a better repair and better documentation of the curing process.

The next several sections will further explain why the above four rules are so important.

Figure 10-62. Illustration of Rule #1.

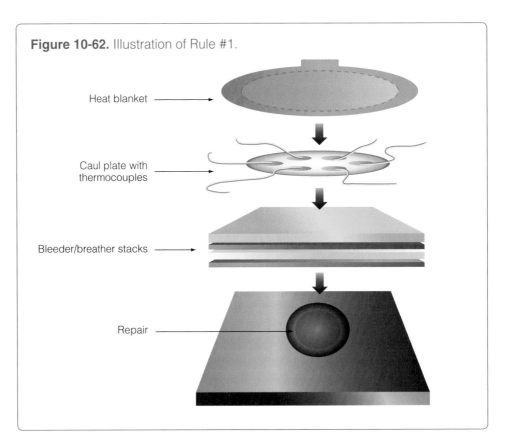

Heat blanket

Caul plate with thermocouples

Bleeder/breather stacks

Repair

Figure 10-63. Illustration of Rule #2.

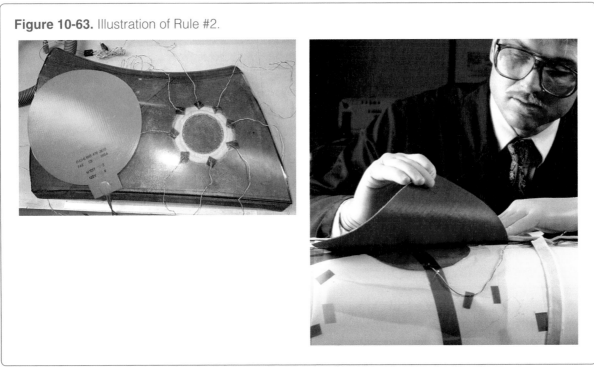

Figure 10-64. Illustration of Rule #4.

The bleeder material is applied first and absorbs excess resin *(top)*, a separator film prevents the resin in the bleeder from saturating the breather, and the breather cloth is applied last, extending over the entire repair area *(bottom)*. Never double up these materials just to fit in the bag.

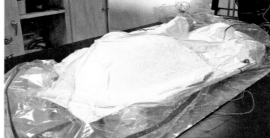

Flash Tape

Put flash tape under the thermocouple to prevent it from making contact with a conductive surface. Then place another flash tape tab on top to secure the thermocouple to the part. A conductive surface could be the composite (fiberglass and aramid composites are not conductive but carbon composites are) or aluminum part being repaired or a metal caul plate. Isolating the thermocouple via flash tape also prevents resin from migrating up the thermocouple wire, minimizing clean-up afterward.

When using a caul plate, the thermocouples are placed directly over the repair on top of the caul plate. Again, put down flash tape first, then the thermocouple, and then another flash tape tab to secure it. The caul plate will separate the thermocouples from the repair and conduct heat from directly above the repair area, giving the best possible readings.

Important: Put flash tape under the thermocouple to avoid spikes in temperature readings. If the thermocouple makes contact with an aluminum caul plate, it introduces a third alloy into the thermocouple's constantan-iron circuit, which may affect the calibration of the instrument and could cause erroneous readings.

Cross-talk

Cross-talk occurs when very low voltage currents develop between thermocouples that are taped directly to a conductive part, without being insulated. The *cross-talk* current causes minor, very rapid fluctuations in the temperature readings, which then could spur the hot bonder to incorrectly add or subtract heat to the repair. This typically results in an alarm being triggered. In the worst case, this can cause a sub-standard repair. In all cases, it will cause alarms to print out on the log records, which then have to be explained to quality assurance and regulatory authorities. The best practice is to prevent cross-talk by insulating thermocouples on any surface, especially electrically conductive surfaces.

Reverse-Wired Thermocouples

It is possible to burn up a part by wiring thermocouples backwards. Connecting the constantan wire to the positive plug (red wire to smaller plug) or the iron wire to the negative or J-plug (white wire to larger plug) will cause reverse polarity in the circuit. In other words, when the temperature goes up in the repair, the hot bonder will read it as going down.

This becomes dangerous very quickly, as some hot bonders might read a decreasing temperature—even though it is actually rising—and so it supplies more heat, which makes the incoming readings look like the part is cooling down even further, so the hot bonder adds more heat, etc., until the part ignites. Most new-generation hot bonders have software and/or hardware protection against this event.

But don't thermocouples come already wired? In some cases, yes, but if they are re-used for multiple repairs they will eventually have frayed and broken wires and will need to be repaired or re-fabricated. Often, new thermocouple wire is purchased in long spools, and it will be up to the technician to cut them to length, weld the working end, and wire on the plugs. Thus, it is important to understand how to wire thermocouples correctly.

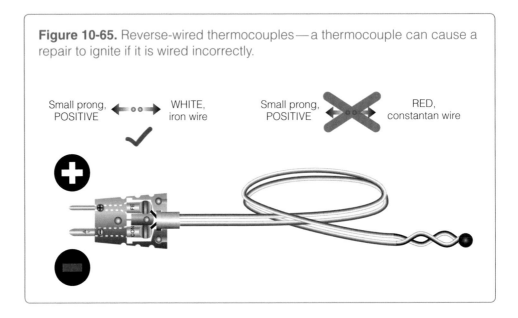

Figure 10-65. Reverse-wired thermocouples—a thermocouple can cause a repair to ignite if it is wired incorrectly.

Caul Plates

Caul plates are smooth metal plates, roughly the same size and shape as a composite repair, used to distribute uniform pressure and temperature across the repair during cure. (*See* Figure 10-66.)

As illustrated in Figure 10-62 (Page 244), caul plates are placed on top of the bleeder/breather stack, and provide a smooth surface for the finished laminate. (*See* Figure 10-67, next page.) They also enable placement of thermocouples closer to the center of the repair.

Caul plates can be difficult to form to complex contours. Also, metal caul plates can cause errors in temperature readings and false alarms as well, unless insulated with flash breaker tape (a.k.a. flash tape). Silicone rubber caul plates do not conduct electrical current like metal plates, but caution still must be taken to prevent thermocouple mark-off.

Thermocouple wires should be bared and welded to form a proper lead on the end that is taped to the composite part. Soldering introduces a third metal or alloy into the circuit (typically constantan-iron) and will cause the same erroneous readings described in the previous sections, "Flash Tape," and "Cross-talk." (*See* Figure 10-68 on the next page.)

Figure 10-66. Caul plates—two views of the same repair show caul plate size and thickness.

Figure 10-67. Caul plates.

This repair begins with a repair patch *(left)*, followed by a sequence of vacuum bagging materials, including bleeder and peel ply, and then a rectangular caul pate *(middle)*. A heat blanket sized larger than the repair area is then applied *(right)*.

Figure 10-68. Insulating thermocouples.

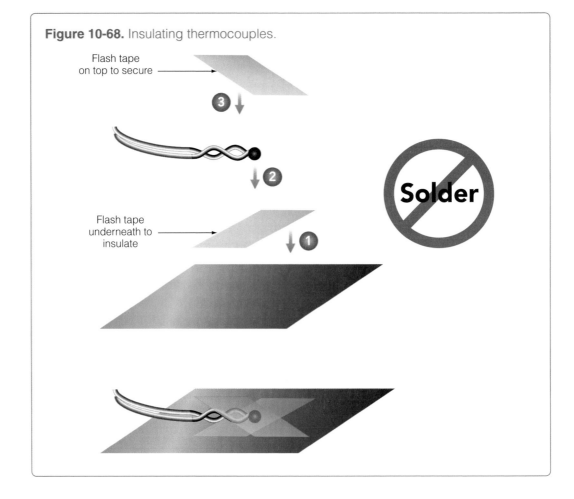

Flash tape on top to secure

Flash tape underneath to insulate

Solder

Health and Safety

11

In this chapter:
- Routes of Exposure
- Hazards Associated with Matrix Systems
- Hazards Associated with Fibers
- Exposure Limits
- Solvents
- Personal Protective Equipment

The subject of safety and advanced composite materials has always sparked tremendous debate. Since accurate data on these materials is often difficult to find, conjecture and hearsay are often deeply rooted in an organization's safety policy where composites are concerned. To compound the problem, an organization may fall under the jurisdiction of more than one occupational health agency, and these agencies may have differing opinions on regulations and priorities.

It is important to understand the different aspects of safety to minimize the level of hazard your employees are being exposed to. It is also important to understand the roles of the industrial hygienist and the safety officer in setting policy that not only complies with the law, but also does not encumber the employee to the point where the quality of the work suffers.

The information contained here is not all-encompassing and should not be regarded as a replacement for safety information provided by the manufacturers of these products (such as a Material Safety Data Sheet, or MSDS). Instead, this section is meant to give a broad overview of composite safety issues such as routes of exposure, and hazards associated with matrix systems and fibers. It will also discuss exposure limits and the role they play in establishing engineering controls, personal protective equipment, and setting organizational policy. Finally, some of the solvents commonly used in the composites industry will be briefly discussed.

Routes of Exposure

Hazardous materials may enter the body in four different ways: absorption, inhalation, injection, and ingestion. Certain materials under the right circumstances may be able to enter the body

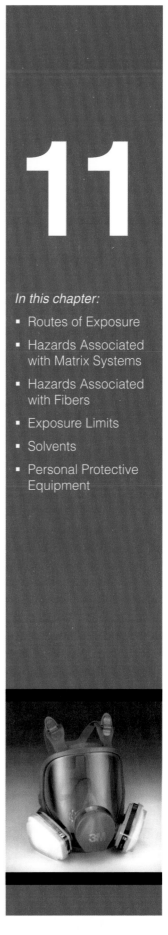

249

via two or more of these routes. Conversely, one route of exposure may be used by two or more constituent materials to enter the body.

Absorption

The skin is a protective barrier that helps keep foreign chemicals out of the body. However, some chemicals can easily pass through the skin and enter the bloodstream. If the skin is cut or cracked, chemicals can penetrate through the skin more readily. Many chemicals, particularly organic solvents, dissolve the oils in the skin, leaving it dry, cracked, and susceptible to infection and absorption of other chemicals. The eye is particularly at risk from the absorption of chemicals as well. Many chemicals may burn or irritate the eye, but they may also be absorbed through the eye or the capillaries in the eyelid and then enter the bloodstream.

Inhalation

The most common type of exposure occurs when you breathe a substance into the lungs. The lungs are made up of branching airways called bronchi. Clusters of tiny air sacs called alveoli are at the ends of these airways. Through the walls of the alveoli, gas exchange takes place. Blood in capillaries outside the alveoli gives up carbon dioxide and takes on oxygen. If other chemicals are present, they too may be absorbed into the bloodstream.

Injection

This method introduces foreign material to the bloodstream or tissue by piercing the skin. It is relatively common in the field of composite repair since damage to structures often results in very sharp, stiff splinters. Wounds caused by splinters of this sort often get infected, but it is important to note that usually contaminants on the splinter cause the infection and not the composite itself.

While limited, the use of hypodermic devices in both composite manufacturing and repair is also a consideration, as the risk of accidental injection of a solvent, resin, or adhesive may exist. This area of risk and hazard is well documented and referenced in the medical community, which may be the best resource for information in this area.

Ingestion

The least common source of exposure in the workplace is swallowing chemicals. Chemicals can be ingested if they are left on hands, clothing or beard, or accidentally contaminate food, drinks, or cigarettes.

Hazards Associated With Matrix Systems

While there are ceramic as well as metal matrix materials, the vast majority of matrix systems used in the composites industry are plastics, or resins. These can be further divided into two categories, thermoset and thermoplastic resins (*see* Chapter 2). Thermoplastics rely on heat to melt, or flow the resin in order to process it, while thermoset resins generally consist of two components that, when mixed and cured properly, form a solid.

Epoxy

The most commonly used thermosets are the epoxy resins. They are made by reacting epichlorohydrin with a suitable backbone chemical, usually a bisphenol A or F resin, to produce diglycidyl ether of bisphenol A or F (DGEBA or DGEBF).

Chemical Irritants

These base resins are found in most epoxy formulations and are generally considered to have a low order of acute toxicity and are only slightly to moderately irritating to the skin. However, some resin formulations containing reactive diluents such as glycidyl ethers can range from moderate to severe as skin and eye irritants. In any case, direct contact with these materials in their liquid form, or their vapors, should be avoided.

Most of the hardeners for epoxies used today contain aromatic or aliphatic amine compounds. Aromatic amines such as methylene-dianiline (MDA), Diaminodiphenylsulfone (DDS), and others are irritants to the eyes, skin, and respiratory tract. Symptoms of overexposure range dramatically from dizziness and headache to liver and possible retinal damage. Some amines, such as MDA, are also classified as Group 2 carcinogens or reasonably anticipated as being carcinogenic.

Like the base epoxy resins, these compounds have a very low vapor pressure and rarely present an airborne hazard unless the mixture is sprayed or cured at high temperatures. However, potential for dermal exposure remains high. Because of the toxicity of this class of hardeners, it is necessary to minimize or avoid exposure by all potential routes.

Several other types of curing agents are also used in the advanced composite industry. These include aliphatic and cycloaliphatic amines, polyaminoamides, amides, and anhydrides. The aliphatic and cycloaliphatic amines, such as diethylene triamine and triethylene tetramine, are strong bases and are usually severe skin and eye irritants. Many of these products can cause skin or respiratory sensitization. While the amide and polyaminoamide hardeners are generally less irritating to the skin, they are still considered potential skin sensitizers. There are two basic types of anhydride curing agents used in today's epoxies. Both types are severe eye irritants as well as strong skin and respiratory tract irritants. As with all materials mentioned in this section, the manufacturer's material safety data sheet (MSDS) should always be consulted prior to use.

Dust

Dust released from the sanding and machining of completely cured epoxy products is generally considered to be "nuisance" dust. The presence of filler materials such as fiberglass, calcium carbonate, powdered silica, lead, etc., may result in concentrations of airborne dust that are more toxic or irritating than nuisance dust. In addition, resins may not be completely cured for days after hardening. Therefore, inhalation of dusts from grinding, sawing, drilling, polishing, etc. of hardened but incompletely cured resins may cause allergic responses in individuals previously sensitized to uncured resins or curing agents.

Sensitization

Repeated exposure to certain constituents in resin systems may result in a condition known as *sensitization*. Sensitization, or allergic contact dermatitis, manifests itself as an allergic skin reaction characterized by red, itchy, irritated skin. In rare, severe cases, people who have been sensitized to the vapors of some constituents may exhibit asthma-like symptoms.

Polyester and Vinyl Ester

Polyester and vinyl ester resins are typically solutions of a solid resin in styrene (or methacrylate or vinyl toluene) monomer and may contain small amounts of other ingredients such as dimethylaniline (DMA), an accelerator, and cobalt naphthenate (CoNap), a promoter. Styrene is an acute eye, skin, and respiratory irritant, and can cause sleepiness and unconsciousness (narcosis) in high concentrations. The resin mixture is a viscous liquid with a strong, distinct odor.

These resins are cured by adding a small amount of an initiator (1 to 3 percent), typically an organic peroxide such as methyl ethyl ketone peroxide (MEKP), or benzoyl peroxide (BPO). These peroxides are strong irritants and can cause severe damage to the skin and eyes. Proper safety precautions must always be observed when using these materials. Additionally, all promoters such as cobalt napthanate should be mixed thoroughly with the base resin before introducing an initiator like MEKP. Under no circumstances should a promoter be mixed directly with an initiator, as violent decomposition and fire may result.

Polyurethane

These systems are formed by reacting a polyether or polyester polyol component with one of three isocyanate compounds, usually toluene diisocyanate (TDI), methylene diisocyanate (MDI), or hexamethylene diisocyanate (HDI). While the polyols generally have a low order of toxicity, the isocyanates are usually severe skin and respiratory irritants. Both skin and respiratory sensitization can readily occur from exposure to isocyanates. Additionally, respiratory sensitization can be caused by skin contact alone.

Polyimide and Bismaleimide

Since these resins have not been studied as extensively as the others, little information exists concerning their potential safety issues. Manufacturer's material safety data sheets indicate that prolonged or repeated contact with these resins may result in skin irritation or sensitization not unlike many epoxy formulations.

Phenolic

This is a family of resins made by reacting one of at least seven different phenols with an aldehyde such as formaldehyde or furfural. Only exposure to phenol and formaldehyde vapors is likely to occur; however, both are potential eye, skin, and respiratory tract irritants. Additionally, phenol can be readily absorbed through the skin in toxic amounts, so both gloves and respiratory protection may be necessary.

Hazards Associated With Fibers

Handling dry carbon fiber, aramid, or fiberglass fabric is not an inherently hazardous activity. The potential to infiltrate and damage the human body is only present once the materials are processed with a matrix material and then machined. Machining a cured composite laminate generates a significant amount of dust. Although the dust created from machining appears to be homogeneous, it actually contains distinct resin and fiber components. The main safety concern that must be addressed is whether or not the particles, especially the fibers, can be considered *respirable*, according to the Occupational Safety and Health Administration (OSHA). Respirable fibers can penetrate deep into the lungs and damage the alveoli. *See* Figure 11-1.

When composites undergo a machining process such as sanding, the relatively fragile matrix material shatters and exposes the fibers. The liberated fibers are then broken off at various lengths by the sanding process. It is

Figure 11-1. Fiber and resin constituents from machining operation.
Because dust from machining contains resin and fiber components, the main safety concern is whether the particles, especially the fibers, are respirable. Respirable fibers can penetrate deep into the lungs and damage the alveoli.

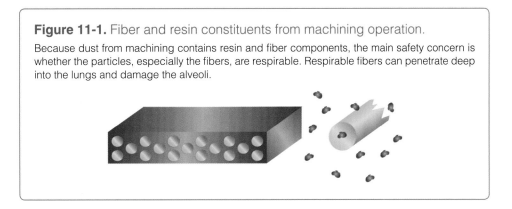

important to note that as the fibers are abraded and broken, the diameter of the fiber remains largely unchanged. Since the vast majority of glass, carbon, and aramid fibers are greater than 6 microns in diameter, the percentage of fibers that fall into the respirable category according to OSHA is usually very small. However, that does not relieve an organization from accurately quantifying all exposures to hazardous materials in the workplace.

Respirable fibers are those whose shape and size allows them to potentially penetrate deep into the lungs and damage the alveoli, the tiny structures in the lung that are responsible for gas exchange. Generally, a fiber is considered respirable if it is 3 microns or less in diameter, less than 5 microns in length, with an aspect ratio greater than 3:1. Fibers larger than this are considered inhalable, but not respirable. Since inhalable fibers are too large to become permanently lodged in the alveoli, they are typically removed by normal lung clearance mechanisms such as coughing.

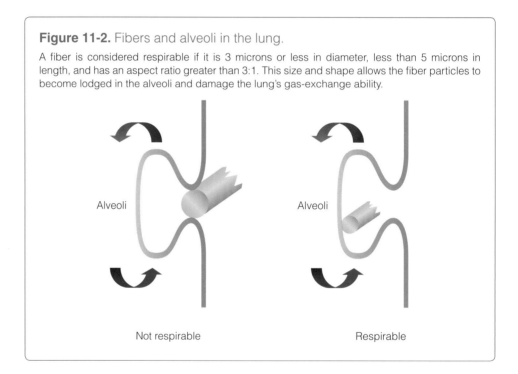

Figure 11-2. Fibers and alveoli in the lung.

A fiber is considered respirable if it is 3 microns or less in diameter, less than 5 microns in length, and has an aspect ratio greater than 3:1. This size and shape allows the fiber particles to become lodged in the alveoli and damage the lung's gas-exchange ability.

Exposure Limits

In the United States, exposure limits are established by OSHA to control exposure to hazardous substances. Collectively known as Permissible Exposure Limits (PELs), they are set forth in OSHA regulations and employers who use regulated substances must control exposures to below the PELs for these substances. PELs are usually expressed in parts per million (PPM) for liquids, gasses or vapors, and in milligrams per cubic meter (mg/m^3) of air for dusts and particulates. Occasionally, airborne fibers are quantified in terms of the number of fibers per cubic centimeter (f/cm^3).

There are three types of OSHA PELs:

1. The 8-Hour Time Weighted Average (TWA) is the average employee exposure over an 8-hour period, based on industrial hygiene monitoring. The measured level may sometimes go above the TWA value, as long as the 8-hour average stays below. All chemicals with PELs have a TWA value. Only a few chemicals have Ceiling and Excursion Limits.

2. The Ceiling Limit is the maximum allowable level. It must never be exceeded, even for an instant.

3. The Excursion Limit is a value that can be exceeded only for a specified short period of time (between 5-15 minutes), which is called the Excursion Duration. When there is an Excursion Limit for a substance, exposure still must never exceed the Ceiling Limit, and the 8-hour average still must remain at or below the TWA.

Industrial Hygiene Reports

In order to determine if a specific procedure or task does or does not generate exposures above the limits set forth by OSHA, a safety officer will usually enlist the help of an industrial hygienist. These specialists possess the knowledge and equipment necessary to quantify exposures in the workplace and will usually generate a report based on their findings. This industrial hygiene report is then used to determine what engineering controls or personal protective equipment should be used in the performance of that particular task.

Threshold Limit Value

Another figure that safety specialists should consider is the Threshold Limit Value, or TLV. This limit is established by the American Conference of Government Industrial Hygienists (ACGIH). It is used to express the degree of exposure to contamination in which most people can work consistently for 8 hours a day, day after day, with no harmful effects.

OSHA PELs are the law concerning the use of hazardous materials. TLVs are recommendations, scientific opinion based solely on health factors; there is no consideration given to economic or technical feasibility of implementing controls to keep worker exposure levels below this level. If they are included on an MSDS, you will notice that TLVs are often lower than PELs for a given substance. While the PEL is the legal standard to which an organization is held responsible, a prudent safety officer might err to the side of caution when determining the necessary engineering controls or personal protective equipment (PPE) to implement in order to control exposure.

Carbon fiber dust is considered by OSHA to be dust NOC (not otherwise classified). As such, its PEL is 15 mg/m^3 total dust, of which 5 mg/m^3 is allowed to be respirable. The ACGIH has assigned carbon fiber dust a TLV of 10 mg/m^3. If an industrial hygiene survey of a particular procedure showed exposure to be approximately 12 mg/m^3, a responsible and prudent safety officer might re-evaluate the engineering controls in place for that procedure to see if the exposure can be reduced below the TLV.

Solvents

Solvents may be used sparingly or extensively depending on the type of components being manufactured and the type of manufacturing process being employed. They might be used at many stages from the impregnation of dry fabrics (prepregging), to the clean-up of tools, parts, and the application of release agents on the molds. Generally speaking, most of the solvents used in the composites industry fall into one of two chemical families: ketones, and chlorinated solvents.

Ketones

The ketone family includes acetone, methyl ethyl ketone and methyl isobutyl ketone. In addition to being extremely flammable, these chemicals pose a significant inhalation hazard. Over-exposure to vapors can cause irritation to the nose and throat, headache, dizziness, nausea, shortness of breath, and vomiting. Higher concentrations may cause central nervous system depression and unconsciousness.

The ketones absorb readily through the skin producing mild to moderate irritation, redness, itching, and pain. Chronic or prolonged skin contact can cause dermatitis and may also cause depression of the central nervous system. Vapors may irritate the eyes and contact with the liquid will produce painful irritation and possibly eye damage. Persons with pre-existing skin disorders, or eye problems, or impaired respiratory function may be more susceptible to the effects of chemicals in this family.

Chlorinated Solvents

The family of chlorinated solvents is quite large containing hundreds of chemicals. Two that are used in the composites industry are trichloroethane (and its variants—trichlorotrifloroethane, and trichloroethylene), and methylene chloride.

Trichloroethane and its variants can be absorbed by intact skin. They cause central nervous system (CNS) depression, and are known to damage the kidneys and liver. Additionally, trichloroethylene is a suspected carcinogen. It can adversely affect the lungs, and can cause rapid and irregular heartbeat that can cause death. Causes skin, eye, and mucous membrane irritation. Toxic effects are increased when combined with alcohol, caffeine, and other drugs. It has been used extensively in vapor degreasers. Many members of this family of chemicals are similar to Halon, and thus displace oxygen. This makes knowing the exact chemical involved critical in choosing a respirator for the task.

Methylene chloride is listed by the ACGIH as a potential carcinogen. As a result, OSHA has recently lowered the PEL for this chemical. Overexposure causes CNS depression, liver and kidney damage, and can cause elevated blood carboxyhemoglobin (also caused by exposure to carbon monoxide). Contact of the liquid with skin or eyes causes an extremely painful irritation similar to a burn. In the aerospace industry, it has long been used as the active ingredient in paint stripper.

Solvents in Release Agents

Exposure to solvent-borne release agents may be one of the most overlooked hazards in the workplace. Engineering controls for proper ventilation and containment of the Volatile Organic Compounds (VOCs) found in these chemistries are paramount to maintaining a safe working environment.

Hydrocarbon solvents used in release agents range from aliphatic to aromatic to chlorinated. At one time chlorinated solvents were very popular because most are not flammable. However, they have been mostly regulated out of use, because they are Ozone Depleting Compounds (ODCs) and suspected carcinogens.

Aromatic hydrocarbons are the most common solvent for release agent systems. They range in flash point from combustible down to flammable. Aromatic solvents are VOCs; they contain Hazardous Air Pollutants (HAPs) and most are suspected carcinogens. To reduce VOCs, manufacturers will use low-density solvents. Lower weight per gallon equals fewer VOCs per gallon. However, these lighter weight solvents are more flammable.

Aliphatic hydrocarbons range from combustible to flammable and are relatively low in VOCs. Many of them are HAP-free and are not known carcinogens. These are usually a safer choice for the end-user; however, non-VOC chemistries may ultimately be the safest and the most environmentally friendly option for today's composite manufacturing factory.

Personal Protective Equipment

When dealing with the chemicals and other hazards in the advanced composite industry, it is important to use Personal Protective Equipment (PPE) that is adequate for the task. The guidance offered in this section is general in nature and not meant to be specific to, or suggestive of the PPE that the reader may require for a given task.

It is important to note that while PPE is important, it is the last in a series of defenses against hazards in the workplace. Exposure to hazardous materials in the workplace should first be mitigated by engineering controls, then administrative controls. If these are impractical, or simply do not reduce the exposure below acceptable limits, then the use of personal protective equipment should be explored.

Engineering controls are apparatus designed to remove hazardous materials as close to their point of origin as possible. This includes equipment such as downdraft tables, exhaust hoods, paint booths, etc.

Administrative controls are rules implemented by an organization to regulate how long any employee may be engaged in activities where they are exposed to hazardous materials.

Personal protective equipment may still be necessary to supplement engineering and administrative controls in order to best protect the operator from the four primary routes of exposure, as outlined below.

Protection from Absorption

Gloves are the most common protection against absorption hazards, worn to protect the hands while handling resins, adhesives, or solvents. The proper glove should be selected based on the materials being handled. For instance, when using a small amount of isopropyl alcohol, nitrile gloves may be appropriate. However, if using acetone, nitrile gloves will break down quickly and expose the operator to the solvent. The actual use of the glove must be considered as well as the degradation and permeation ratings of the glove material. Always consult a reputable supplier of safety equipment before selecting gloves for a specific task.

Since the category of absorption includes exposure of the eyes, safety glasses, goggles, or face shields may also be required if the materials are prone to splashing. In that case, other items of protective clothing such as plastic aprons or even Tyvek^{TM} suits may be warranted. Much like glove selection, the material selected for this protective gear is determined by the hazards involved.

Protection from Inhalation

For respiratory hazards, the primary protection should be derived from engineering controls in the work area. Ventilation hoods, exhaust fans, etc. may be available that can bring exposure levels down below permissible exposure limits. This is not only the preferred solution according to OSHA, but in the long run, it is likely to be the most economical because it is a one-time expense.

If engineering controls are not sufficient, are impractical, or are unavailable, then appropriate respirators are necessary. Respirators are a continuing expense that not only includes the equipment, but also the recurrent training for end-users. This may involve medical examinations, screening, and regular health monitoring of the employees. Establishing a respiratory protection program can be relatively simple or quite complex, depending on your workplace and the number of people involved. Your local safety equipment vendor will likely have a list of companies in your area who are qualified to conduct this sort of training. Whether your workplace is large or small, it is important that respirators be used in compliance with OSHA regulations (in particular, 1910.134) and in accordance with the equipment's NIOSH approvals.

Figure 11-3. Respirators protect against inhalation. (Photos courtesy of 3M)

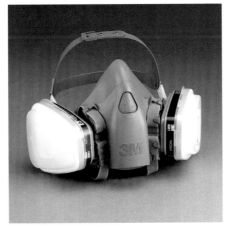

3M™ Half Facepiece 7500 Series Respirator   3M™ Full Facepiece 6000 Series Respirator

Limiting the Risk of Injection

Protection against injection is difficult because the type of equipment that would be needed to guard against a sharp object penetration tends to be stiff and difficult to work in. New materials and technologies in protective gear are making this more attainable. Obviously, the best protection is not rushing the work, and paying close attention to the task at hand. Proper first aid supplies and training are another aspect of protection in this category. As mentioned earlier, infection is as big a concern as actual chemical hazards in this mode of injury.

Preventing Ingestion

We all know better than to eat or drink the chemicals we work with in this industry. What we don't always think of is the trace amounts that might be on our hands when we grab a quick snack in the workplace, or the little bit that is on our fingers if we have a cigarette at break time.

The main two ways to protect against this method of entry is good personal hygiene (washing hands before eating/drinking/smoking) and the use of gloves. If gloves are worn and then removed before taking a break, it is that much less likely that chemical hazards will be ingested.

Glossary

For readers' convenience, some of the term entries in this glossary are highlighted in bold blue type where they occur in the main text, in order to facilitate lookup whenever an unfamiliar term may be encountered. This list represents just a few of the terms and definitions used daily by the advanced composites industries.

3D fabrics. *See* three-dimensional fabrics.

accelerator. A material that, when mixed with a catalyst or a resin, will speed up the chemical reaction between the catalyst or hardener and the resin.

adhesive. A substance capable of holding materials together by attachment to each surface. Structural adhesives are those which are capable of transferring considerable structural loads between materials.

adhesive (or liquid) shim. Typically, paste adhesive used as a gap-filling material between two mechanically fastened substrates.

adhesively bonded. Two or more materials that are attached to each other with an adhesive, typically resulting in a permanent bond.

advanced composites. Generally considered to be made of fibers with lower density, higher strength or modulus than E-glass, combined with a matrix resin that has significantly better thermal and mechanical properties than unsaturated polyester or vinyl ester resins.

amide curing agents. Basically, ammonia with a hydrogen atom replaced by a carbon/oxygen and organic group; used as a hardener for epoxy resins.

amine curing agents. Basically, ammonia with one or more hydrogen atoms replaced by organic groups. Aliphatic amines and polyamines are common curing agents for epoxy resins.

anhydride curing agents. Agents such as Dodecenyl Succinic Anhydride (DDSA), Methyl Hexahydro Phthalic Anhydride (MHHPA), Nadic Methyl Anhydride (NMA), and Hexahydrophthalic Anhydride (HHPA). They provide for good electrical and structural properties and exhibit a long working time. Anhydrides require an elevated temperature cure.

amorphous carbon. A non-crystalline form of carbon. often related to coal or coke carbon foam products.

aramid fiber. A structural fiber made of aromatic polyamide, often recognized by the brand name Kevlar®.

A-scan. A non-destructive ultrasonic pulse-echo inspection method where ultrasonic sound waves are transmitted and received by a single transducer through a composite laminate.

A-stage prepreg. Early stage in the reaction of a thermosetting resin in which the resin is still soluble and fusible. (*See also* B-stage and C-stage prepregs)

ASTM. American Society for Testing and Materials; an internationally recognized standards organization.

asymptote/asymptotic. In this context, a line or curve with very little change in shape, usually related to a vertical or horizontal axis.

atmospheric pressure. The pressure exerted on all things by the weight of the earth's atmosphere. The accepted standard-day pressure at sea level is 29.92 inches (760 mm) of mercury, or 14.7 psi. This pressure varies inversely with altitude and decreases by approximately $\frac{1}{2}$ pound for every 1,000 feet in altitude gained.

autoclave. A pressure vessel used for processing composite materials at pressures exceeding one atmosphere. Typical autoclaves will have heating, cooling, pressurization, and vacuum/vent systems. They are usually pressurized with an inert gas such as nitrogen.

Automated Ply Verification (APV) device. A high-resolution camera designed to work with a laser projector to verify and document fiber alignment compared to preset fiber axial tolerances. Patented by Assembly Guidance Systems, Inc.

Automated Fiber Placement (AFP). An automated fiber placement system where the mold or mandrel moves on multiple axes as well as the robotic fiber placement head. The machine lays fiber bundles in engineered axial patterns against the mold or mandrel, to produce a composite layup.

Automated Tape Layup (ATL). An automated tape layup system. An automated tape layup head, usually on a moving gantry, is used to lay up unidirectional tape widths on a fixed mold or mandrel in an automated tape laying process.

balanced laminate (in-plane). A laminate that has an equal number of "minus" and "plus" angled plies for torsional equilibrium.

basket weave fabric. Fabrics with interwoven multiple yarns, bundled side by side (for bulk) and woven in a plain weave format. (*See also* woven roving)

bending stiffness. The relative stiffness of a beam or plate of a material when a bending moment is applied, usually measured with a three-point bending apparatus. Resistance to out-of-plane flexure.

benzoyl peroxide (BPO). A chemical in the organic peroxide family, consisting of two benzoyl groups joined by a peroxide group. Typically used as a catalyst for reacting polyester or vinyl ester resin systems.

biaxial-stitched reinforcement. A fabric consisting of one layer of unidirectional tows stitched to another layer of unidirectional tows, in which each layer has a different fiber-axis angle; typically 0/90 or +45/-45.

bismaleimide (BMI). A type of thermoset polyimide resin that cures by an addition reaction rather than condensation reaction, thus minimizing volatiles, and therefore void formations during cure. BMIs are tougher, and serve a higher temperature range than epoxies.

bisphenol A. Condensation product formed by the reaction of phenol with acetone. This polyhydric phenol is the standard intermediary that is then reacted with epichlorohydrin in the formulation of epoxy resins.

bisphenol F. Condensation product formed by the reaction of phenol with formaldehyde, which is then reacted with epichlorohydrin in the formulation of epoxy resins. (Also known as a novolac epoxy formulation.)

blank. "Blank" is a broad description used to identify any material form that is to be used for a subsequent process. It can be anything from a dry preform to a pre-consolidated stack of thermoplastic sheets awaiting the next step in a cutting, kitting, or molding process. (*See also* charge)

blasting. The process of using alumina, sand, plastic, wheat starch, or any other media through a high pressure nozzle to abrade the surface of a substrate for the purpose of paint or coating removal, or surface preparation for painting or bonding.

bleeder. An absorbent layer used in the vacuum bag schedule to allow resin to move or "bleed" into the layer during processing. Typically used in non-autoclave processes.

bond. To adhere to a material and provide substantial chemical and/or mechanical adhesion to the material at the bonded interface.

bond assembly fixture (BAF). A tool used to locate and bond materials and/or components together in a primary, secondary, or co-bond process.

bondline. The line of adhesive between two substrates.

bond strength. The stability of a bond, usually described in terms of bond energy, which is the energy required to "fail" the bond, or the amount of energy that the bonded joint can take without failure.

boron fiber. A high-modulus fiber made by vapor deposition of boron gas onto tungsten filament, producing a strong, stiff fiber.

bow-tie coupon. A tensile or compressive test coupon resembling a bow tie, that is narrow at the mid section and wider at the grip ends to facilitate failure in the mid-section of the specimen. (*See also* dog bone coupon)

braid, braiding. Braiding intertwines three or more yarns so that no two yarns are twisted around one another. In practical terms, "braid" refers to fabrics continuously woven on the bias. The braiding process incorporates axial yarns between woven bias yarns, but without crimping the axial yarns in the weaving process. Thus, braided materials combine the properties of both filament winding and weaving, and are used in a variety of applications, especially in tube construction.

breather. A breather layer is used in the vacuum bag schedule to maintain a pathway for air and gas to escape from the processing laminate to the port or vents within the bag.

bridging. Refers to a condition where the fibers in a laminate span or "bridge" across a radius or curve in a part, as a result of a lack of contact or compaction to the previous layer during layup or processing of the panel.

B-scan. An ultrasonic inspection technique rarely used on composite materials because it sends an ultrasonic sound wave along the laminate, in-plane, along the width or length of the laminate (rather than through the cross-section).

B-stage prepreg. An intermediate cure stage of a thermosetting resin that is between completely uncured and completely cured. All thermosetting prepregs are supplied in a B-stage condition. The degree of B-staging will advance as a prepreg material ages. (*See also* A-stage and C-stage prepregs)

buckling. Compressive deformation of fibers typically combined with matrix failure in a laminated panel, beam, or channel.

bulking materials. Typically, a non-woven core such as Soric® used within the laminate to create bulk or thickness.

bulk molding compound (BMC). Thermoset resin mixed with short fiber reinforcements and fillers into a viscous compound for use in compression molding processes.

carbon fiber. Carbonized polyacrylonitrile (PAN) or pitch-based fiber with higher stiffness (flexural or tensile modulus) than that of glass and many other fibers. Also referred to as graphite fiber.

carbon matrix/carbon-carbon composites (CCC). Carbon-carbon composites consist of carbon fibers embedded in a carbon matrix. Typically fabricated by building up a carbonized matrix on a fiber preform through a series of resin impregnation and pyrolysis steps.

carbonize/carbonization. The process of pyrolyzing a precursor material to a carbon form, usually at temperatures around 1,315°C (2,400°F), in an inert atmosphere. Actual carbonization temperature is dependent on the precursor materials, manufacturer's specifications, and the desired material properties of the end-product.

catalyst. A substance that changes the rate of a chemical reaction without itself undergoing permanent change in composition or becoming a part of the molecular structure of the product. It markedly speeds up the cure of a compound when added in minor quantity as compared to the amounts of primary reactants.

caul plate. A smooth metallic, elastomeric, or composite sheet or plate used on the bag-side of a composite layup or repair to provide a smooth, wrinkle-free surface to a localized area of the finished laminate.

centipoise. A measurement of viscosity; 1 centipoise is 1/100 Poise, equal to the viscosity of water. (*See also* viscosity)

ceramic fiber. Continuous fibers made from metal oxides or refractory oxides which are resistant to high temperatures (2,000–3,000°F / 1,093–1,649°C). This class of fibers includes alumina, beryllia, boron, magnesia, zirconia, silicon carbide, quartz and high silica (glass) reinforcements. Ceramic fibers are created by chemical vapor deposition, melt drawing, spinning and extrusion. Their main advantage is high strength and modulus. (Although glass and carbon fibers are also ceramic materials, they are not generally included in this list.)

ceramic matrix composites (CMC). Ceramic matrix composites combine ceramic reinforcing fibers with a ceramic matrix to create superior high-temperature properties.

CFRP. Carbon fiber reinforced plastic.

charge. A "charge" is typically a tailored or cut piece of material made from a larger blank, which is subsequently loaded onto a mandrel or into a mold for processing. (*See also* blank.)

chemical vapor deposition. Precursor gases such as elemental boron are introduced into a reaction chamber at ambient temperature. As they come into contact with the heated substrate, (such as a tungsten filament in the manufacture of boron fiber), they decompose, forming a solid material which is deposited onto the substrate.

chopped strand mat. A material form made with randomly placed chopped strand, typically in a tissue or mat form, designated by the fiber type and areal weight.

clockwise (CW) symbol. Refers to the fiber orientation symbol, a mirror image of the counterclockwise symbol used on composite manufacturing drawings to designate the fiber orientation for production of the composite part. Useful in repair instructions where applicable.

closed molding. Any molding process in which Volatile Organic Compounds (VOCs) are either contained or controlled from release into the atmosphere.

closed cell foam. Any solid foam product that has a closed cell structure where the material is not porous between cells in the structure.

cobond, cobonding. The process of bonding a precured detail to an uncured laminate, requiring surface preparation and the use of an adhesive at the bondline. Cobonded composite repairs fall under this definition.

cocure, cocuring, cocured. The practice of curing all active thermoset materials in one process.

coefficient of thermal expansion (CTE). The change in length or volume per unit length or volume produced by a 1° rise in temperature.

cold mixed resin. In this context, comparatively, a "cold" mixed batch of polyester or vinyl ester resin could be mixed with less peroxide so that it reacts slowly and provides more working time. With epoxy resins, a cold mix using less hardener is not an option, as epoxies require precise measurements of each polymer component to achieve proper structural properties.

combination fixture, *or* jig. A tool designed to provide more than one tool or operational function.

compaction of prepreg, *or* laminate. Refers to interim or continuous compaction of combined resin and fiber forms in the mold, upon the mandrel, or in-situ at the machine head in an automated material placement process.

compaction shoe. A mechanical device used at the head of an automated material placement machine that provides in-situ compaction pressure to the material-form being applied in the automated process.

composite. A composite is a material comprised of two or more individual materials in which each retains its own unique properties, but when combined, the resulting material has superior properties to that of its constituents.

compression molding. A molding process using matched cavity dies mounted in a press, where bulk molding compound (BMC) is introduced into the open die set and is compressed at high pressure and temperature while being redistributed throughout the cavity as the press/die is closed. The die remains closed at elevated temperature for the duration of the cure cycle.

compressive strength properties. The ability of a material to resist a force that tends to crush or buckle; the maximum compressive load sustained by a specimen divided by the original cross-sectional area of the specimen.

compressive stress. The normal stress caused by forces directed toward the plane upon which they act; the compressive load per unit area of original cross-section carried by the specimen during the compression tests.

computer controlled processing. Controlling process parameters using a computer software program and a computer control interface to a PID Loop Controller (PLC.)

constituent. A single element in a larger grouping. For example, in advanced composites, the principal constituents of a composite laminate are the fibers and the matrix resin.

contact molding. A spray-up or layup process where resin and fiber are placed in a mold or on a mandrel, rolled or squeegeed in place; but no additional pressure is applied to the layup in the process as it cures, typically at room temperature.

contamination. An impurity or extraneous substance in a material, on a surface, or in the environment that affects one or more properties of the material, for example, adhesion.

continuous fiber reinforcement. Refers to a fiber form or fabric where the fibers run continuously throughout the form of the material (as opposed to chopped or discontinuous fiber reinforcements).

continuous strand mat. Non-woven reinforcement made with randomly oriented continuous fibers. They are typically used as surface ply materials in structural laminates and as reinforcements in semi-structural applications.

controller. Equipment used to control temperature, pressure, vacuum, or other aspects related to composite processing.

copolymer. A long-chain molecule formed by the reaction of two or more dissimilar monomers.

core forming jig, *or* fixture. A tool specifically used for forming a core material to a desired shape using a vacuum bag, press, or other mechanical clamping mechanism at an elevated process temperature.

core potting compound. A syntactic paste used to pot core close-outs, core ramps, hard-points, etc.

core splice, core splice adhesive. Edgewise splice of core materials, typically using a foaming core-splice adhesive at butt-joints or by overlapping 2–4 honeycomb core cells, using a "crush core" process where the overlapped core cells are driven or "staked" together.

counterclockwise (CCW) symbol. This fiber orientation symbol is shown from the manufacturing point of view, where the fiber orientation is viewed from the inside of the panel looking toward the tool surface (as would be the case during layup). The symbol controls the primary fiber orientation where it is shown on the structure on the drawing.

couplant. A liquid or gel-like material used to "couple" to a solid surface to allow sound waves to travel from the transducer, through the couplant to facilitate an ultrasonic inspection.

cosmetic repair. A repair to a scratch, dent, or a ding of a size and/or location that is accomplished using filler putty or other non-structural material simply to restore the appearance of the part.

course (in stitched fabrics). Number of stitches of yarn per inch in a stitched fabric in the longitudinal/warp direction.

covalent bond. A chemical bond formed by sharing one or more electrons, particularly pairs of electrons, between atoms.

creel. A tension-controlled spool, designed to control the release of fiber from an attached spool of tow, usually arranged in a multi-creel array, supporting an automated fiber placement or filament winding operation.

cross-linking. A process in which polymer molecules or chains are chemically attached or "linked" to one another, resulting in a single complex polymeric molecule. Refers to a reacting thermoset resin system.

C-scan. A through-transmission ultrasound inspection technique.

C-stage prepreg. The final stage of the evolution of a thermosetting resin in which the material has become infusible and insoluble in common solvents. Fully vitrified thermosets are in this stage. (*See also* A-stage and B-stage prepregs)

cure, curing (or, fully reacted). To permanently change the properties of a thermosetting resin system as a result of a controlled chemical reaction, usually involving heat and pressure. This is a nonreversible process as the resin goes from a viscous or liquid state to a solid or elastic state.

cure cycle. Typically refers to a time/temperature recipe prescribed to cure a thermoset resin system. Technically, it is the time and temperature required to achieve the desired properties of a thermoset resin.

cure temperature. The temperature required to vitrify or fully cure a thermoset resin system.

cure time. The time required to vitrify or fully cure a thermoset resin system.

curing agent. The hardener or catalyst that activates the chemical reaction in a thermoset resin system.

cyanate ester. A high performance thermoset resin characterized as having a low coefficient of thermal expansion, good hot-wet properties, and a low dielectric constant.

cycle time (or, production time). The total time that it takes to process (heat, soak, and cool) a part or load of parts in a process vessel such as an oven or autoclave.

damage tolerance. The resistance of a component or structure to failure when damaged.

debulking, vacuum debulking. Intermittent compaction steps during a layup of a composite panel to achieve good consolidation of materials, especially in a complex shape. Typically this is done under a vacuum bag for a short period of time. A longer debulk may be prescribed to remove gas and air and further compact a laminate prior to final processing.

decitex. Unit of weight measure of strand or yarn in grams per 10,000 meters in length.

delamination. The term used to describe a disbond between layers or plies in a laminate.

denier. Unit of weight measure of strand or yarn in grams per 9,000 meters in length.

desiccant. A substance that promotes drying (for example, calcium oxide).

die, die set (or, heated die, matched die). A tool used for forming sheet material in a press, or a die used for forming a shape in a pultrusion process. Also used to define tools used for injection molding, RTM, and compression molding processes (the latter are closed-cavity dies).

dielectric constant. Dielectric means a nonconductor of electricity; therefore dielectric constant is the measurement of the ability of a material to resist the flow of an electrical current.

differential scanning calorimetry (DSC). A thermo-analytical method measuring the amount of energy absorbed (endotherm) or produced (exotherm) during the cure of a thermoset resin system.

disbond. Refers to a lack of a bond between two substrates. Typically, a composite skin to a core disbonds in a sandwich panel.

dog bone coupon. A tensile or compressive test coupon resembling a bow tie, which is narrow at the mid-section and wider at the grip ends in order to facilitate failure in the mid-section of the specimen. (*See* bow-tie coupon)

double bias reinforcement. A fabric form, typically stitched, in which two layers of fibers are oriented at equal angles (i.e., +45° and -45°).

drape, drapeability, drapable. The ability of a fabric to drape or form over a complex shape without cutting and tailoring.

drape formed, drape forming. A process where many layers of prepreg or bulk fiber reinforced thermoplastic materials are heated and formed over a mandrel.

drill fixture, *or* jig. A tool used for drilling holes in a metallic or composite part, typically designed with slip-renewable bushings to minimize wear.

dry pack. Refers to a dry pack or preform of materials placed in a mold, vacuum bagged, and infused in a resin infusion process.

dry T_g. The initial thermal glass transition temperature (T_g) of a cured thermoset resin or adhesive, prior to exposure to moisture or hot-wet conditions.

dynamic mechanical analysis (DMA). Measures the stiffness and damping properties of a material, typically through a change in temperature. DMA can be used to measure the viscoelastic properties of a curing thermoset resin during the cure cycle. Also useful in determining the T_g.

Dyneema® fiber. An ultra-lightweight, high-strength thermoplastic fiber made from ultra-high molecular weight polyethylene (UHMWPE), also known as high modulus polyethylene (HMPE) or high performance polyethylene (HPPE). Dyneema® was developed by the Dutch company DSM in 1979.

edge closure. The edge closure design for a sandwich panel. Design methods range from simple potted edges to tapered close-outs. Also plastic or metallic channels may be used for edge closure design.

edge distance. The distance from the center of a fastener to the net-edge of a panel. Fastener pitch is the distance between fasteners in the same row, measured from center-to-center. Transverse pitch is the distance between rows of fasteners in the same joint.

electromagnetic (EMF) shielding. The process of limiting the penetration of electromagnetic fields into a space, by blocking them with a barrier made of conductive material. Typically it is applied to enclosures, separating electrical devices from outside access. Electromagnetic shielding is also used to block radio frequencies (also known as RF shielding) and electromagnetic radiation.

elongation (to break). Measured amount of elongation of a material loaded in tension to failure. Typically measured with an extensiometer or strain gauge.

end grain balsa. Balsa wood core material cut in sheets where the grain runs perpendicular to the plane of the sheet, exhibiting excellent compressive properties.

emissions. In this context, emissions are volatile organic compounds (VOC) that are emitted into the atmosphere.

encapsulate. To surround, encase, or protect, in or as if in a capsule (such as fibers encapsulated in a plastic matrix).

environmental conditions. Refers to the environment that the structure will be exposed to in service.

epoxy. A high performance thermoset resin that has good thermal, structural, and adhesive properties.

exothermic reaction. A common chemical reaction producing a slight increase in temperature when a thermoset resin evolves during cross-linking. Large masses of resin can produce an out-of-control exotherm, which can cause excessive heat and toxic smoke.

expanded aluminum, or copper. Refers to a metallic "mesh" used for lightning strike protection or EMF/EMI protection.

extraction (of volatiles, air, and gas). The inherent removal of trapped air and gas when the composite laminate is vacuum bagged and "debulked" for a period of time. Routes for extraction are provided in the vacuum bag schedule for debulking.

face skin. The face sheet or skin on a cored sandwich panel or structure. Also known as "facing."

fastener footprint. The area covered by the head of a fastener in a composite or metallic panel. Typically composite panels require a larger footprint than metallic structures.

fatigue. The failure or decay of mechanical properties after repeated applications of stress; the ability of a material to resist the development of cracks, which eventually bring about failure as a result of a large number of cycles. Failure is also the event leading to fracture under repeated or variable stresses having a maximum value less than the tensile strength of the material. Fatigue fractures are progressive, beginning as miniature cracks that grow under the action of the fluctuating stress. In composites, the effect of cyclic damage (fatigue) is different than metals, i.e., splitting and delamination, matrix cracking, and fiber breakage.

fatigue life. The number of cycles of deformation required to produce failure of the test specimen under oscillating (stress or strain) conditions.

fiber angle callout. Primary fiber angles for each ply are controlled by the ply orientation symbol and the ply table callouts on the composite drawing.

fiber buckling. Refers to buckling of reinforcement fibers in a laminate loaded edgewise, on-axis in compression beyond the strength of the matrix that is binding the fibers together.

fiber-dominated properties. The subset of physical properties that are dependent on the type of fiber used in a laminate versus the type of matrix. Tensile strength and flexural stiffness (modulus) are examples of fiber-dominated properties.

fiber placement. See Automated Fiber Placement (AFP).

fiber reinforcement. Any fiber used for the purpose of reinforcing a matrix material to gain composite properties.

fiberglass. A common term that refers to fibers made of glass. (Fibers made of silica compositions, *see* glass fiber.)

fiber volume, fiber volume fraction. The amount of fiber in a composite laminate; usually expressed as a percentage volume fraction.

filament. Describes the basic structural fibrous element. It is either extruded or spun from the precursor materials that are used to make the specific type of fiber desired.

filament winding. A process of continuous fiber placement on a mandrel where both helical and hoop wraps produce the desired fiber axial strength and stiffness.

fill face. Refers to the fill-yarn dominant surface of a harness satin weave fabric. (Opposite of warp face.)

filled resin. Any resin that has been filled with any of a number of fillers such as silica or milled fibers. (*See* syntactic foam)

fillers. In this context, fillers are physical additives such as silica microballoons or milled fibers, used to thicken or enhance other properties within the resin or adhesive.

fill yarn (or, weft yarn). Yarns that run across or cross-wise to the length of a roll of woven cloth.

film adhesive. Flat, "B-staged" structural adhesive with a backing film applied. Typically furnished in rolls not unlike prepreg fabrics. Often used with other prepreg materials or along the bondline in an adhesively bonded structural joint.

finish. A coating applied to a yarn or fabric that provides chemical compatibility with the fiber so that the matrix resin will easily wet and distribute along the fiber during processing.

flame/flammability resistance. The ability of a material to resist burning and releasing heat or flame and toxic smoke.

flame treatment. Refers to using a flame to treat a surface to change the surface free energy prior to bonding.

flash breaker tape. A heat resistant tape used to secure ancillary materials within a vacuum bag for processing. The term "flash breaker" refers to the ability to seal off a layer of film at a constant distance from the edge of a laminate to contain the amount of resin flash around the periphery of a laminate during processing.

flat braid. Flat braids use only one set of yarns, and each yarn in the set is interwoven with every other yarn in the set in a zigzag pattern from edge to edge.

Flex Core®. A Hexcel trademarked core design that allows for the core to flex or bend over complex contours.

flow, in resin. The ability of a resin to flow when processed. This is particularly important to prepreg materials that are approaching or have exceeded their initial out-time.

flexural properties. Flexural stiffness, in this context, is related to on and off- axis flexural loading of a composite laminate.

fracture toughness. The ability of a material to resist fracture when loaded or impacted. Also refers to the measure of damage tolerance of a material containing initial flaws or cracks.

fusion bonding. The ability to fuse two thermoplastic materials together either through a solvent process or by welding. (*See* thermoplastic welding)

gage section (gage length). The part of the specimen over which strain or deformation is measured. (*See* bow-tie coupon)

galvanic corrosion. Corrosion due to galvanic activity between two dissimilar materials, where one is anodic and one is cathodic. The material on the anodic side of the scale experiences the corrosion.

gauge. Number of stitches of yarn per inch in a stitched fabric in the transverse/fill direction.

gauge pressure. Gauge measurement of positive pressure, typically expressed numerically as psig or barg. Gauge pressure is not absolute pressure, which starts at 0 absolute (below atmospheric pressure).

gel point, gelation, gelled. Measured change in viscosity of a thermoset resin by means of rheometric analysis. Typically the gel point is designated at half way up the transition slope before the viscosity peaks at vitrification.

gel coat, gel coated. A surface coat of filled resin used at the face of a laminate for cosmetic reasons (color, etc.), or for void filling purposes in a composite tooling application.

GFRP. Glass fiber reinforced plastic.

glass fiber. Fibers made from one or more formulations containing silica, of which the resulting spun filament has excellent strength to weight, as well as electrical and thermal insulation properties.

glass mat thermoplastic (GMT). Glass fiber reinforced thermoplastic sheet used for compression and press molding processes.

glass transition temperature (T_g). The temperature at which a material goes through a transition from a glassy-solid state to a rubbery, flexible state (or vice-versa).

graphite fiber. "Graphite" is used in pencils, as a lubricant, and in electrodes. Carbon atoms are strongly bonded together into graphite sheets (like mica flake). The term is mistakenly used to describe high modulus carbon fiber, as they are atomically the same but extremely different in form; carbon fibers are manufactured in tension, aligning the molecular structure, providing for a functional fiber that is not as brittle or soft as graphite. Although there is a distinct difference in the form, the industry still refers to carbon fiber as graphite.

hardener. A material added to a polymeric mixture to promote or control the curing action by becoming an integral part of the cured resin.

harness satin (HS) weave fabrics. In this weave patter, a fill yarn floats over a number of warp yarns and "hooks" under one yarn in the fabric, and then repeats the pattern. With all harness-satin weaves, there is a dominant warp and fill face to consider in layup, which effects symmetry.

high modulus carbon fiber. Carbon fibers that have a tensile modulus greater than 50 msi/345 GPa.

high temperature. Generally considered to be temperatures greater than 212°F/100°C, or the boiling point of water at sea level.

heat distortion temperature (HDT). The temperature at which a standardized test specimen deflects a specific amount under a stated load. Also called deflection temperature under load (DTUL).

HOBE (HOneycomb Before Expansion). The HOBE (HOneycomb Before Expansion) is the block of bonded sheets before they are expanded into honeycomb patterns. When made from paper they are expanded and dipped in resin (typically phenolic). Metallic HOBE can be expanded and made into sheets or sold in HOBE form for ease of machining.

honeycomb nodes (or, node lines). These are the bonded interfaces between the honeycomb cells, denoting the "ribbon" direction.

honeycomb ribbon direction. The line, which is parallel to the bonded nodes in the honeycomb core, called out and oriented on the composite panel drawing.

honeycomb web. The wall of the honeycomb cell; much like mini I-beams placed throughout the core sandwich panel.

hot melt prepreg. Prepreg manufactured with heated resin typically coated on a drum and then rolled onto a backing paper prior to the dry fabric being impregnated into the resin in the process. Hot-melt prepregs have little to no volatile content as a result of the process.

hot mix. Comparatively, a "hot" mixed batch of polyester or vinyl ester resin could be mixed with more peroxide so that it reacts quickly and shortens the working time. With epoxy resins, a hot mix using more hardener is not an option, as epoxies require precise measurements of each polymer component to achieve the proper structural properties.

hot-wet service conditions, hot-wet properties. Referring to hot, humid (including salt spray) environments to which composite structures are exposed, and the ability to resist these conditions.

hybrid fabrics. Fabrics woven with two or more different fiber types, such as carbon and aramid fiber hybrid fabric.

hybrid laminates. Laminates with layers of different fiber types such as a carbon fiber laminate with an aramid fiber outer layer in a laminate used for abrasion or impact resistance.

impact resistance. The ability of a material to resist damage from impact. Toughened thermoset resins or thermoplastic resins provide good impact resistance in a composite structure.

infrared radiation. Relating to the invisible part of the electromagnetic spectrum with wavelengths longer than those of visible red light but shorter than those of microwaves. Infrared radiation from the sun is measured as heat.

Infusion, resin infusion, vacuum infusion. A fabrication process that involves the introduction of liquid resin into a dry-pack or fiber preform. (*See* resin infusion)

inhibitor. A chemical added to a liquefied monomer to inhibit polymerization or solidification.

initiator. Technically not a catalyst, but often referred to as such, the initiator starts the reaction that allows the chemistry to evolve to a cured solid. The most commonly used materials for this purpose are Benzoyl Peroxide (BPO) and Methyl Ethyl Ketone Peroxide (MEKP).

initiated resin systems. Technically all thermosets are initiated, however in this context, polyester and vinyl ester resin systems are initiated with a peroxide initiator.

in-plane shear strength. Resistance to in-plane surface distortion of a material.

interlaminar shear. Through-the-thickness distortion of a material.

intermediate modulus carbon fiber. Carbon fibers that have a tensile modulus greater than 42 msi/290 GPa, and less than 50 msi/345 GPa.

Invar (tooling). Invar is an iron-nickel alloy with very low thermal expansion properties. Typically, Invar 36 or 42 (36 or 42% nickel) is used for tooling.

isothermal. Relating to a process at a constant temperature.

Kevlar®. DuPont's brand name for aramid fiber.

kitting. Refers to pre-cutting and packaging "kits" of materials to be used to manufacture a composite component.

lagging thermocouple. Referring to a single thermocouple out of a group of thermocouples, which lags in response to temperature changes when compared to the rest of the group.

lamina. Refers to a single layer or ply of material.

laminae. Plural of lamina, refers to more than one layer or ply over another.

laminate. Refers to multiple layers or plies bonded together; i.e., composite laminate or laminate panel.

laminating resins. Typically, liquid resins that have relatively low viscosity and surface tension values that allow for ease of wetting (of fibers) in a laminating process.

lap shear strength. The calculated strength of a bonded lap joint when the substrates are pulled in tension. Typically lap shear test coupons utilize a $\frac{1}{2}$ inch long x 1 inch wide lap joint, and the breaking strength is multiplied by 2 to provide pounds per square inch values. These values can be used for material qualification but are not applicable to actual lap shear values for use in design.

laser positioning. Refers to the use of a laser projector to outline ply locations or other details in layup or assembly.

laser shearography. Shearography is a nondestructive test using an image-shearing interferometer to detect and measure local out-of-plane deformation on the test part surface, as small as 3 nanometers, in response to a change in the applied engineered load.

layup. The act of laying-up a single ply or layer at a time on a mold, plate, or mandrel. The resulting laminate may also be called a layup.

layup assembly approach to manufacturing. A cocure manufacturing approach where all of the panel layers, stiffeners, etc. are laid up and compacted on separate molds or mandrels and then assembled in the primary mold for final cure.

layup mold or mandrel. A tool used to layup and process the composite laminate, panel, cocured or cobonded structure.

legacy cure cycle. A cure profile or recipe based on legacy time and temperature specifications rather than actual material state feedback processing.

lightning strike protection (LSP). A metallic or conductive surface on a composite structure to dissipate lightning strike energy along the outer surfaces of the structure in order to minimize damage to the structure. (*See* expanded aluminum or copper)

long-fiber reinforcement. A fiber form or fabric where the longitudinal fibers in the roll, bolt, or spool runs full length (as compared to chopped or discontinuous fiber reinforcements). Suited for structural applications where long-fiber load transfer is required.

Long-Fiber Reinforced Thermoplastic (LFRT). Thermoplastic prepregged material containing long fiber reinforcements, typically unidirectional tape used for tape placement or other fabrication processes.

longitudinal properties. Properties along the primary axis of load in a composite panel or structure.

low observable (L.O.). Engineering and material technologies designed to reduce a craft's signature in the visual, radar, sonar, infrared, and audio ranges. Often referred to as "stealth," these technologies are used primarily to enhance the survivability of military aircraft and watercraft.

low pressure molding compounds (LPMC). Low pressure sheet molding compounds (SMC) that process in the range of 150 psi.

low-energy impact. Impact to a structure where the impacted area and subsequent damage may be difficult to detect.

material state conditions, material state management of cure. Material state conditions refer to the actual viscoelastic state of the material being processed. Material state management refers to controlling the cure cycle based on the actual viscoelastic properties of the material, as opposed to using antiquated legacy cure methods.

mandrel. A tool used to layup and process the composite laminate, panel, cocured or cobonded structure.

mat. Any non-woven tissue, veil, or mat made with random short or continuous fibers in a thin sheet form.

matched die forming. A press-forming process using a set of dies (a cavity die-set) to control the inner and outer surfaces of a reinforced thermoset or thermoplastic composite part.

matrix, matrix resin, polymer matrix. A material used to encapsulate or bond fibers together in a fiber reinforced composite structure.

matrix degradation. Degradation of the matrix material, typically from exposure to excessive heat or from chemical attack.

matrix digestion test method (or, acid digestion). Used to remove the matrix material from a laminate specimen in order to interpret the fiber-resin-void volume of a test panel or laminate coupon. (ASTM D3171 standard test method.)

matrix dominated properties. The subset of physical properties that are dependent on the type of matrix used in a laminate versus the type of fiber. Examples are service temperature, interlaminar shear strength, and compressive strength.

matrix-fiber interface. The interfacial bond of the matrix to fiber at the interface between fibers or layers of fibers in a laminate.

matrix ignition loss test method (or, burn-out). Used to remove the matrix material from a laminate specimen in order to interpret the fiber-resin-void volume of a test panel or laminate coupon. (ASTM D2584 standard test method.)

mechanical fasteners. Screws, bolts, blind bolts, rivets, or a number of different mechanical fasteners designed to be used to fasten together a structure.

mechanical properties. Compressive, tensile strength, and modulus of materials associated with elastic and inelastic reaction when force or load is applied; the individual relationship between stress and strain. (*See* structural properties)

mechanically fastened. Refers to a fastened joint where all loads through the joint are transferred through mechanical fasteners.

meta-aramid. A type of aramid fiber that inherently has a non-aligned molecular chain structure. It is very resistant to high temperatures and is used as a substitute for asbestos in industrial applications. It is good for use in making fire-resistant racing suits, and aramid paper for honeycomb core materials. However, it is not a good chemistry for making structural fibers. First developed and patented in the mid-1960s by DuPont and sold under the trade name Nomex™.

metal matrix composites (MMC). Composites containing non-metallic (ceramic) fibers bound with a metal-matrix; usually aluminum, titanium, magnesium or copper.

methacrylate. Refers to any of several plastic substances formed by polymerizing esters of methacrylic acid to make an adhesive or sealant.

methylene chloride. A solvent that will attack and degrade most plastic materials. It is a nonflammable liquid typically used as a paint remover for metallic structures.

Methyl Ethyl Ketone Peroxide (MEKP). Highly reactive organic peroxide used to produce a chemical reaction with a promoter such as cobalt napthenate to start the polymerization of a polyester, vinyl ester, or acrylic resin system, causing it to evolve to a solid and cross-link with a functional monomer to produce a cured thermoset polymer.

microballoons. Typically micro-fine precipitated silica filler (Cab-O-Sil or Visco-Fill) used to thicken a resin system so that it will fill a gap or hang on a vertical surface. Microballoons are non-uniform in size.

microbeads. Usually silica beads manufactured to a very precise diameter, used as spacers along a bondline to maintain a constant bondline thickness.

microcracking. Microscopic cracking of the matrix material between fibers primarily caused by thermal stress fracture, but may also be related to flex or impact stress.

mid-plane. The mid-section at the center (cross-section) of a laminate. Laminates are normally designed to be symmetric about the mid-plane to ensure dimensional stability when exposed to a change in temperature.

mix ratio. The ratio of resin to hardener. (*See* parts by weight *and* parts by volume)

mold (or, tool). A layup mold, mandrel, or any of many different tools used in the production of composite parts. (*See* tooling)

modulus, tensile modulus. Modulus of elasticity is the ratio of stress or applied load to the strain or deformation produced in a material. Modulus of rigidity is the ratio of shearing stress to angular deformation within the elastic region for shear or torsional loading.

moisture contamination of prepreg. Usually moisture contamination occurs when the moisture barrier is removed from the frozen prepreg roll prior to the roll thawing to room temperature. When this happens, condensation brings moisture to the exposed prepreg.

moisture ingression/intrusion. The movements of water or other fluid contaminants into a composite structure, particularly into honeycomb sandwich structure.

monocoque. A type of construction in which the skin or outer shell bears all or most of the load or stress on the structure.

multiaxial fabrics. Typically, a stitched fabric made up of long fibers running in three or more axis of orientation.

natural fibers. Fibers such as sisal, hemp, flax, cotton, etc. used in the production of composite parts.

neat-resin. Resin as to which nothing (fibers, fillers, etc.) has been added. Usually the base resin for further formulation.

nesting. Orientation of harness-satin weave plies so the faying or mating surfaces "nest" or have the dominant (warp or fill) fibers running on the same axis of orientation at the interface.

Non-Destructive Testing (NDT), Non-Destructive Inspection (NDI). One of many different test or inspection techniques that do not require destruction of a specimen or part.

non-woven core material. A lightweight, non-woven material typically used for the dual purpose of both a core and an interlaminate resin distribution media for a resin infusion process.

non-woven materials. A compressed tissue or mat made up of non-woven short or continuous fibers that are bound to each other mechanically or with a binder. Also applies to stitched materials.

oil-canning. In this context, refers to the ability of an asymmetric laminate, that when loaded from one side, it curls or warps about the dominant axis on the loaded side of the panel. When loaded from the opposite side of the panel, the laminate "flips" and curls along the opposite axis, making a noise similar to that of an old-fashioned oil can.

open cell foam. Any solid foam product that inherently has an open cell structure where the material is porous between cells in the structure.

open time. The time that a resin or adhesive is left open and exposed to the atmosphere before is covered or joined.

open molding, open mold. A mold that controls a single surface. Today, an open mold is typically used for vacuum bagged, oven or autoclave cured composite parts, however some non-vacuum bag open molding is still done in the industry.

operational service temperature. A service temperature of a composite that is set conservatively lower than the wet T_g of a thermoset or thermoplastic resin matrix. (*See* service temperature)

out-of-plane shear. Shear stress effected through the cross-section of a composite laminate. Typically tested via a short beam shear test. (ASTM D2344 standard test.)

out-time. The accumulative time that a prepreg is out of the freezer. Limited to a specified amount of time.

oven, oven cure. A non-pressurized process vessel capable of heating. Process ovens normally have ramp programmable controllers for processing (curing) a variety of materials.

overbraid. This is a technique where fibers are braided directly onto cores or tools that will be placed within molding processes. It is often used when a preform with very high bias angles or a contoured triaxial design is required, or when it is desirable to include circumferential windings in a preforms architecture.

PAN. Polyacrylonitrile. Used as a precursor for making carbon fiber. (*See* polyacrylonitrile)

para-aramid. A type of aramid fiber that has a highly aligned molecular chain structure, making it a good material for making structural fibers. First developed in 1973, patented by DuPont and sold under the trade name Kevlar®.

paste adhesive. A thermoset adhesive that usually contains fillers, making it paste-like, allowing for gap filling and the ability to hang on a vertical surface.

P.B.V. (parts by volume). Refers to a resin mix ratio expressed in parts by volume.

P.B.W. (parts by weight). Refers to a resin mix ratio expressed in parts by weight.

peel ply. A layer of fabric, either porous or non-porous, applied to the backside of a laminate to provide for a protective layer that peels away from the laminate upon removal.

permeability. The rate of passage of a gas, liquid, or a solid through a barrier without chemically or physically affecting it.

phased array inspection. An ultrasonic inspection technique using an array transducer that has many elements, which allows for a variety of wide area and localized point inspection capabilities within a composite panel.

phenolic. A thermosetting resin that is produced by condensation of an aromatic alcohol with an aldehyde (phenol with formaldehyde). Phenolic functions at higher temperatures than epoxies; however, it is generally considered to be a low strength resin.

pitch carbon fiber. Carbon fiber made from petroleum pitch through a high temperature production process, resulting in intermediate, high, and ultra-high modulus carbon fibers for use in structural applications.

plain weave. Consists of yarns interlaced in an alternating fashion, one over and one under every other yarn. The plain weave fabric provides stability, but is generally the least pliable, and least strong, due to a lower yarn count and a higher number of crimps than other weave styles. Plain weave fabrics tend to have significant open areas at the numerous yarn intersections. These intersections will be either resin rich pockets or voids in a laminate made with this fabric style. There is no warp- or fill-dominant face on a plain weave fabric.

plasma treatment. Plasma surface treatment is used to raise the surface free energy of a plastic surface in preparation for bonding.

pleats (in a vacuum bag). Pleats or "dog ears" are placed along the periphery of a vacuum bag and sealed with sealant tape. This allows for plenty of bag materials for pleating at the base of core transitions, or changes in geometry within the bagged part area.

ply layup table. The ply layup table is the master tabulation of ply numbing, sequencing, orientation, material callout, splice requirements, symbol relevance, and revision information located on the face of the composite panel drawing.

ply orientation, ply angle callout. Refers to the alignment of the fibers of each layer in the laminate, on the desired axis, inspected at the point designated by the location of the universal ply orientation symbol and the ply table on the drawing.

ply orientation shorthand. Refers to ply orientation shorthand code, typically used in engineering and design for communication of complex laminate layups in a simplified manner.

ply orientation symbol. The primary orientation symbol on the plan view of the composite panel that controls the fiber axial direction relative to the symbol, applicable only where the symbol is located on the panel. If there are multiple symbols used on the drawing, then that information is coordinated with the ply table.

Poise. A measurement of viscosity. From *Poiseuille* (named for the French physician, Jean Louis Poiseuille, who researched flow resistance of liquids extensively in 1846). The common unit for expressing absolute viscosity is *centipoise* (1/100 Poise, 1/1000 of a Poiseuille).

polyacrylonitrile (PAN). A synthetic resin prepared by the polymerization of acrylonitrile. An acrylic resin, it is a hard, rigid thermoplastic material that is resistant to most solvents and chemicals and of low permeability to gases. Most polyacrylonitrile is produced as acrylic and modacrylic fiber used as a precursor for making carbon fiber.

polyamide (PA). A thermoplastic polymer containing repeated amide groups. A type of condensation polymer produced by the interaction of an amino group of one molecule and a carboxylic acid group of another molecule to give a protein-like structure. The polyamide chains are linked together by hydrogen bonding.

polyamideimide (PAI). A thermoplastic amorphous polymer that has exceptional mechanical, thermal and chemical resistant properties. Polyamide-imides are produced by Solvay Advanced Polymers under the trademark Torlon®.

polycarbonate (PC). Polycarbonates have high impact strength and exceptional clarity. These unique properties have resulted in applications such as ballistically resistant windows and shatter resistant lenses, etc. More recently however, low flammability of polycarbonate is of interest to industry. This material is formed by a condensation polymerization reaction of bisphenol A and phosgene.

polyester resin (unsaturated). A thermoset resin made of unsaturated polyester mixed with a vinyl-active monomer such as styrene. The cure is affected through polymerization using a peroxide catalyst along with a promoter and an accelerator. The end result is a cross-linked polymer (to the styrene monomer).

polyester resin, (saturated) thermoplastic (PET). Polyethylene terephthalate (PET) is a polymer resin from the polyester family and is used in synthetic fibers, thermoforming applications, and engineering applications often in combination with glass fiber. (*See* polyethylene terephthalate).

polyetheretherketone (PEEK). PEEK is a crystalline thermoplastic, which makes it a very chemically resistant product. PEEK exhibits excellent chemical resistance including organic solvents, aqueous reagents. The heat distortion temperature for PEEK is 600° F/316°C, with continuous service above 450°F/232°C. PEEK has high mechanical properties, very low smoke emission and self extinguishing characteristics. Therefore it is used for many demanding applications, for example: automotive engine parts, aerospace components, compressor valve parts, etc.

polyetherketoneketone (PEKK). PEKK is a high-temperature thermoplastic that has high stiffness and strength and excellent chemical resistance. The term PEKK actually represents a variety of copolymers with different ratios of terephthalate (T) and isophthalate (I) moieties. By varying the T/I ratio, it is possible to control the crystallization rate and ultimate crystallinity and mechanical properties of the polymer, without substantially reducing its end-use temperature. Despite its many attractive properties, the commercial application of PEKK has been limited by its high cost and its relatively poor melt processability.

polyetherimide (PEI). PEI is an amorphous, amber-to-transparent thermoplastic with similar characteristics to PEEK. Relative to PEEK, it is cheaper, but less temperature-resistant, and lower in impact strength. It is prone to stress cracking in chlorinated solvents. Available under the trade name Ultem®, a registered trademark of SABIC Innovative Plastics.

polyethylene (PE). Any of the polymers of ethylene, i.e. high-density and ultrahigh-molecular-weight polyethylene (HDPE and UHMWPE, respectively), or branched to a greater or lesser degree (low-density and linear low-density polyethylene; LDPE and LLDPE, respectively). The branched polyethylenes have similar structural characteristics (e.g., low crystalline content), properties (high flexibility), and uses. HDPE has a dense, highly crystalline structure of high strength and moderate stiffness. UHMWPE is made with molecular weights 6–12 times that of HDPE; it can be spun and stretched or aligned into stiff, highly crystalline fiber with a tensile strength many times that of steel. Uses include bulletproof vests and protective armor.

polyethylene terephthalate (PET). A saturated, thermoplastic, polyester resin made by condensing ethylene glycol and terephthalic acid. It is extremely hard, wear-resistant, dimensionally stable, resistant to chemicals, and has good dielectric properties.

polyimide (PI) thermoset. Polyimides include condensation and addition polymers. Addition-type polyimides based on reacting maleic anhydride and 4,4'-methylenedianiline are processable by conventional thermoset transfer and compression molding, film casting and solution fiber techniques.

polyimide (PI) thermoplastic. A synthetic polymeric resin originally developed by DuPont that is very durable, easy to machine and can handle very high temperatures. Polyimide is also highly insulative and does not outgas. Vespel® and Kapton® are examples of thermoplastic polyimide products from DuPont.

polymer. A high molecular weight organic compound, natural or synthetic, typically a thermoset or thermoplastic polymer, or elastomeric material.

polymer matrix composites. Composites made with plastic matrix materials, either thermoset or thermoplastic.

polymerization. A chemical reaction in which the molecules of a monomer are linked together to form large molecules whose molecular weight is a multiple of that of the original substance. When two or more monomers are involved, the process is called copolymerization.

polyphenylene sulfide (PPS). PPS is a high-strength, highly crystalline engineering plastic that exhibits good thermal stability and chemical resistance. It is polymerized by reacting dichlorobenzene monomers with sodium sulfide at about 480°F/250°C in a high-boiling, polar solvent. It is used for many fiber-reinforced parts such as battery boxes and covers, etc.

polypropylene (PP). Polypropylene (PP) shares some of the properties of polyethylene, but it is stiffer, has a higher melting temperature, and is slightly more oxidation-sensitive. A large proportion goes into the production of fibers, where it is a major constituent in fabrics used in industry.

polysulfone (PSU). Polysulfone (PSU) is a rigid, high-strength, semi-tough thermoplastic that has a heat deflection temperature of 345°F (174°C), and maintains its properties over a wide temperature range.

polyurethane. A thermoset resin made from the reaction of diisocyanates with polyols, polyamides, alkyd polymers, or polyether polymers.

porosity. In this context, porosity refers to surface voids in a laminate.

post-cure. A post cure is necessary when a thermoset resin has been vitrified at a temperature below the target service temperature. Therefore it is "post-cured" at temperature intervals or at a very slow rate so as to "push" the T_g to higher temperatures without allowing distortion of the matrix, typically during a free-standing post-cure.

pot life. The time at which a freshly mixed thermoset resin has at 25°C (77°F), in a known amount, at a designated thickness, before it starts to gel or rapidly change viscosity (not necessarily the working time).

potting compound. A lightweight syntactic or filled resin used to "pot" or fill the edges or other hard-point locations in honeycomb core to prevent core crushing during processing.

precursor, carbon fiber. A precursor refers to the material used as the primary ingredient in the manufacture of a product. In the case of carbon fibers, pitch and PAN are precursor materials. (*See* pitch *and* PAN)

preform. Preforms are fiber reinforcements that have been oriented and pre-shaped to provide the desired contour and load path requirements of the finished composite structure. They are most commonly used in resin transfer molding (RTM), where their complex fiber orientations provide multiple flow paths for the injected resin, enabling good fiber encapsulation and higher properties.

prepreg. A reinforcement material that is pre-impregnated with resin. The reinforcement can be made from any type of fiber, in a unidirectional, stitched, or woven form. With thermosets, the resin is already mixed prior to saturating the reinforcement. The amount of resin in the prepreg is strictly controlled to achieve the desired resin content. It is stored in a moisture barrier and kept frozen until needed to preserve out-time.

press. A hydraulically-driven, heated set of platens, or special set of heated dies. It has high-pressure capabilities used for processing composites and other materials.

pressure (or, equilibrium pressure). Refers to when the hydraulic pressure of the resin in a vacuum bag comes to equilibrium with the atmospheric pressure, causing a loss of down-force or compaction on the underlying laminate.

promoter, promoted resin systems. Typically refers to polyester or vinyl ester resins that have had the promoter (Cobalt Napthenate) added to the resin formula at the manufacturer. This eliminates the hazard of storing and mixing three-part systems.

primary structure. Also known as a "fracture critical" structure that carries or transfers primary loads throughout a major, heavily loaded structure. Failure of a primary structure typically is catastrophic. (*See* secondary structure)

primer. Typically, a corrosion-inhibiting layer sprayed or otherwise applied to a metallic surface prior to painting or structural adhesive bonding. The term also applies to an interface coat of resin used for wetting a surface of a composite prior to applying a paste adhesive.

processing. The act of following a process to achieve consistent results; for example, processing thermoset resins.

production rate. The number of parts or assemblies being manufactured within a certain amount of time.

pulse-echo ultrasonic inspection. Also known as an A-scan, a non-destructive, ultrasonic pulse-echo inspection method where ultrasonic frequency sound waves are transmitted through, and received by, a single transducer through a composite laminate. (*See* A-scan)

pultrusion. An automated production process in which all of the raw materials are pulled in tension through a die or die-set in order to form a shape. The shape is cured as it exits the pultrusion machine and is cut to length.

quadraxial reinforcement. A four-layer stitched fabric, usually with one layer in each of the four primary directions: 0°/+45°/90°/-45°.

quartz fiber. Quartz fiber is produced from quartz crystal, which is the purest form of silica. The crystals, which are mined mainly in Brazil, are ground and purified to enhance chemical purity. The quartz powder is then fused into silica rods, which are drawn into fibers and coated with size for protection during further processing. Quartz fiber has the best dielectric constant and loss tangent factor among all mineral fibers. For this reason, along with its low density (2.2 g/cm^3), zero moisture absorption, and high mechanical properties, quartz fiber has been used frequently in high-performance aerospace and defense radomes. Quartz fiber also has almost zero coefficient of thermal expansion in both radial and axial directions, giving it excellent dimensional stability under thermal cycling, and resistance to thermal shock.

quasi-isotropic. Refers to the in-plane isotropic properties achieved in laminates using multi-axial orientations of the fibers to attain similar properties when the laminate is loaded in-plane, in flexure, compression or tension, on any axis.

R-value. Refers to a material's resistance to heat transfer. The higher the "R" value, the better the insulation properties.

ramp rate. The rate at which the temperature is rising or falling in a ramp cycle. (*See* ramp cycle)

ramp/soak cycle. The portion of a cure recipe in which the temperature is ramping up, holding at a specified temperature, and cooling down. (*See* cure cycle)

rate of cure. Refers to the time compared to temperature required to cure a thermoset resin at elevated temperatures; the "rate of cure" (or rate of chemical reaction) is accomplished faster at temperatures above the glass transition temperature than below it. Therefore, final cure temperatures are typically engineered to be as near as practical to the final desired T_g in order to minimize the cure time or rate of cure.

recertification/requalification. Refers to materials that have exceeded their shelf life but may be deemed usable through testing. Usually prepreg materials will be recertified or requalified for use after a flow/gel test has been conducted to verify that the resin still meets original certification parameters. The material then is recertified for a short duration and again can be used in production of components.

release film. A low-energy thermoplastic film such as Teflon® or some other low-energy polymer film used in the vacuum bag schedule for the purpose of releasing from the resin, blocking, or restricting resin flow.

release agents/release systems. Semi-permanent polymer coatings that are used on the layup tool or mandrel to provide release from the tool after processing. Also refers to polyvinyl alcohol (PVA) and paste wax used in tooling and other room-temperature composite manufacturing operations.

resin. A solid or pseudo-solid organic material, generally of a high molecular weight. In fiber reinforced plastics, it is the resin that binds the fibers.

resin chemistry. Refers to the specific chemical makeup of a selected resin or resin system.

resin content (or, matrix content). The amount of resin in a composite laminate, usually expressed as a percentage by weight or volume.

resin/fiber ratio. The amount of resin to fiber expressed as a percentage of each; i.e., 40/60 resin to fiber by volume or by weight.

resin infusion. A fabrication process that involves the introduction of liquid resin into a dry-pack or fiber preform.

resin pyrolization. The carbonation of a resin matrix, usually phenolic, by ultra-high temperature exposure, in the manufacture of carbon-carbon composites.

resin transfer molding (RTM). RTM is a process in which mixed resin is injected into a closed mold containing fiber reinforcement. It produces a part with good surface finish on both sides and is well-suited to complex shapes and large surface areas.

reticulated film adhesive. According to Webster's Dictionary, *reticulation* is defined as "resembling or forming a net or network." Therefore, reticulated adhesive forms a network of fillets around the honeycomb core nodes when it is heated during processing.

rheometer. A torsion plate rheometer is used to measures the stiffness and damping properties of a composite material through a change in temperature. A rheometer can be used to measure the viscoelastic properties of a curing thermoset resin during the cure cycle. Also useful in determining the T_g. (*See* DMA)

ribbon direction. Refers to honeycomb core; the direction of the bonded nodes between layers in the manufacture of core.

room temperature cure. A thermoset resin system that cures at "room temperature" (77°F/25°C); or, the time it takes a thermoset resin to cure to desired properties at room temperature.

roving. A roving is a number of strands, yarns, or tows collected into a parallel bundle with little or no twist. Single-end rovings are smaller in diameter and are more expensive to produce due to the higher precision process required for the smaller diameter. Multi-end rovings are larger in diameter and generally less expensive, but can be more difficult to process in filament winding and pultrusion applications. Rovings are most commonly used for constructing heavy fabrics. (*See* woven roving)

sanding ply or layer. A sacrificial ply, typically applied to the outer layer of a repair, designed to allow for post-process sanding of the repair to blend or "feather" the repair to a suitable finish prior to primer and paint application.

sandwich construction. Sandwich panel laminates are designed with an integral lightweight core material to provide web space between laminate skins, increasing the stiffness and strength-to-weight ratio of the panel. Sandwich panels are processed at lower process pressure than solid laminate panels to prevent core crushing. Typically, sandwich laminate panels are less damage-tolerant than solid laminate structures.

scarf. A tapered or otherwise-formed end on each of the pieces to be assembled using a scarf joint.

scarf joint, scarf joint repair, scarfed repair, scarfed structural repair. A joint design that consists of a uniformly tapered angle machined through a composite laminate. This tapered angle is matched in a corresponding laminate that is then secondary-bonded in place, or matched ply-for-ply with equivalent layers in a cobond process to create a very robust bond joint in the composite structure. (*See* tapered or scarf joint)

sealant, sealer, sealing system. Using an adhesive to seal around fasteners, external doublers, and various repairs in order to prevent liquid or gas ingress into a structure. Usually a polysulfide or polythioether based material, but many different materials can be used for this purpose.

secondarily bonded, secondary bonding. In secondary bonding, two or more already-cured composite parts are joined together by the process of adhesive bonding, during which the only chemical or thermal reaction is the curing of the adhesive itself.

secondary structure. Generally considered to be less "fracture critical" than a primary structure, yet still carries significant loads throughout a structure. Failure of a secondary structure typically does not lead to catastrophic failure. (*See* primary structure)

semi-preg. A material that has been partially impregnated with a resin layer, typically on one side of a dry fabric.

semi-rigid tooling. Usually refers to localized tooling (cauls) made with a combination of flexible and rigid materials. These tools are frequently used to control steep core-ramp angles, detail interface locations, and edgeband smoothness along the bag-side surface of a cocured composite panel.

separator layer/film, in vacuum bagging. The separator layer is used between the bleeder layer and the subsequent breather layer to restrict or prevent resin flow. This is usually a solid or perforated release film that extends to the edge of the layup, but stops slightly inside the edge of the bleeder layer, to allow a gas path to the vacuum ports.

service environment. The environmental conditions to which a composite structure is exposed during its service life.

service temperature. The service temperature of a composite is set at a temperature that is conservatively lower than the wet T_g of a thermoset or thermoplastic resin matrix.

service-life. The designed or actual service life of a structure.

S-2 glass. "S" glass refers to structural grade glass with higher tensile strength and modulus than that of "E" glass. Pure S glass is no longer manufactured and S-2 glass has replaced it industry-wide as it has similar properties and is less expensive to manufacture.

shear properties. Properties related to shear stress, strain, modulus, etc., resulting from applied forces that cause two adjacent parts of a bonded laminate or joint to slide relative to each other in the direction parallel to their plane of contact. In interlaminar shear for example, the plane of contact is mainly resin.

sheet molding compound (SMC). A fiber reinforced thermosetting compound, manufactured into sheet form, rolled up into rolls separated with a plastic backing film and used to manufacture a variety of different composite parts and panels.

shelf life. The amount of time that materials such as prepreg, liquid resins, or adhesives have at their specified storage temperature and still meet specifications or maintain suitable function.

shrinkage. The relative change in dimension from the length measured on the mold at room temperature, compared to the length of the molded object at room temperature 24 hours after it has been removed from the mold.

signal attenuation. The diminution of a sound signal over time or distance, such as when performing ultrasonic inspection on composite materials.

silane finish. Silane is a known couplant agent for polyester, vinyl ester, and epoxy resins. It is used as a finish on glass fibers/fabrics to better wet the fiber and facilitate bonding. (*See* finish)

size (applied to fibers). A chemical coating applied to fibers during initial manufacturing. Size is typically added at 0.5 to 2.0 percent by weight and may include lubricants, binders and/or coupling agents. The lubricants help to protect the filaments from abrading and breaking. Coupling agents tend to cause the fiber to have an affinity for an individual resin chemistry. *Binder*, *size* and *sizing* are often used interchangeably in the industry, *size* is the correct term for the coating that is applied, and *sizing* is the process used to apply it.

smoke toxicity. Composite materials produce smoke when they burn that consists of a potentially toxic mixture of gases, airborne fibers, soot particles, etc. The degree to which it is toxic is determined through testing and analysis of the specific material.

soak. The isothermal hold for a specified period of time in a cure recipe designed to allow for a "full cure" of a thermoset resin system.

solid laminate. A "solid" laminate made of all fiber-reinforced resin layers and no added core material. Distinguished from sandwich panel construction, solid-laminate panels can be processed at much higher pressure than sandwich panel laminates as there is no risk of crushing a core material. Typically, solid laminate panels are more damage-tolerant than sandwich structures.

solvent. A substance, usually a liquid, capable of dissolving another substance; i.e. acetone, alcohol, methyl ethyl, ketone (MEK), methylene chloride, etc.

Spectra® fiber. An ultra lightweight, high-strength thermoplastic fiber made from ultra high molecular weight polyethylene (UHMWPE), also known as high modulus polyethylene (HMPE) or high performance polyethylene (HPPE).

splice. A butt-splice, overlap, or other specific splice joint may be required for the fabrication or repair of composite structures. Typically splice requirements are controlled using a process specification. These specifications will be noted on the associated part or assembly drawing.

stacking. In this context, this term refers to "stacking" harness satin (HS) weave fabrics as opposed to "nesting" these materials. A stacked laminate has different performance characteristics than a nested laminate. (*See* nesting)

stacking sequence. The engineered sequence in which the layers or plies of a laminate are arranged to meet design performance requirements.

state of cure. The actual material state condition of a thermoset resin during a cure cycle. Typically measured with analytical equipment reading real-time sensor data, parallel to the actual cure of a composite panel.

static-dissipative. Refers to the ability of a material to conduct or dissipate static electricity to ground. Measured in ohms, this classification refers to a material with an inherent resistivity range between 1×10^4 ohms and 1×10^{11} ohms.

stealth. *See* low observable.

stiffness-to-weight ratio. Comparative stiffness to weight of different materials; also known as the specific stiffness.

strand. A number of filaments bundled together to form a strand; for example, a single glass strand is comprised of anywhere from 51 to 1,624 filaments.

strength-to-weight ratio. Comparative strength to weight of different materials; also known as the specific strength.

structural properties. Compressive, tensile strength, and modulus of materials associated with elastic and inelastic reaction when force or load is applied; the individual relationship between stress and strain. (*See* mechanical properties)

styrene. A colorless aromatic liquid that can be easily polymerized, which is used in organic mixtures, specifically in the manufacture of synthetic rubber and plastics.

surface coat. A thin layer of filled or unfilled resin applied to the surface of a tool or part laminate for cosmetic reasons.

Surface Free Energy Value (SFEV). Denotes the measurement of the total energy content of a solid surface in equivalence to the surface tension of a liquid. Values are expressed in SI units of measure: mN/m and mJ/m^2.

Surface Tension Value (STV). Denotes the measurement of the total energy content of a liquid in equivalence to the surface free energy of a solid surface. Values are expressed in SI units of measure: mN/m and mJ/m^2.

symmetry. Refers to symmetry of the fiber axial orientations about the mid-plane of a laminate. Laminate symmetry is required for dimensional accuracy and elimination of in-plane and out-of-plane behavioral coupling.

syntactic foam. Compounds made by mixing hollow microspheres of glass, epoxy, phenolic, etc. into fluid resin systems to form a lightweight, formable paste. (*See* filled resin)

tack. Referring to the stickiness, or the lack thereof, of a resin or prepreg material.

tapered or scarf angle. The angle of taper of a scarf joint; i.e., the taper ratio of length to thickness.

tapered or scarf diameter. The outer diameter of a circular shaped tapered or scarf repair removal area. This corresponds to the diameter of the largest *original* repair ply.

tapered or scarf distance. The calculated distance from the edge of the removed damage to the outermost edge of the tapered angle.

tapered or scarf joint. A joint design that consists of a uniformly tapered angle machined through a composite laminate. This tapered angle is matched in a corresponding laminate, which is then secondary-bonded in place, or matched ply-for-ply with equivalent layers in a cobond process in order to create a very robust bond joint in the composite structure.

tap testing. A simple non-destructive test involving a coin or small tap-hammer used to lightly tap the surface of a composite laminate and listen for subtle changes in the auditory response, indicating the possibility of an inclusion, delamination, or disbond in the area.

tensile properties. The measured tensile stress and strain (and related calculated properties, such as modulus) of a given material.

tex. A unit for expressing linear density equal to the mass or weight, in grams, of 1,000 meter increments of filament, strand, yarn, or other linear fiber forms.

TFNP. Teflon® Fabric Non-Porous.

TFP. Teflon® Fabric Porous.

T$_g$. *See* glass transition temperature.

thermal analysis (TA). Any one of many analytical techniques developed to examine physical or chemical changes of a sample of material that occurs as the temperature of the sample is increased or decreased.

thermal conductivity. The ability of a material to conduct heat between two surfaces of a given material with a temperature difference between them. The thermal conductivity is the heat energy transferred per unit time and per unit surface area, divided by the temperature difference. It is measured in watts per degree Kelvin.

thermal cycle. Refers to a cycle where the material experiences a change in temperature from ambient to a specified temperature (or set of temperatures), eventually returning to ambient.

thermal expansion. The change in dimension related to a change in temperature for a given material.

thermal imaging (thermography). A non-destructive inspection method involving the use of an infrared camera in order to look at the thermal signature of a panel or structure, in an effort to locate defects or anomalies within the panel or structure.

thermocouple. A temperature measurement sensor made up of two dissimilar wires where a change in temperature at the welded or joined end creates a change in voltage, which is measured and converted into a temperature reading.

thermocouple mark-off. The molded imprint of thermocouple wires when they are run across the back side of a laminate during fabrication or repair.

thermoform, thermoforming. The processes of heating and forming a sheet (or pre-consolidated stack) of thermoplastic materials across a mold or mandrel.

thermo mechanical analysis (TMA). Thermo mechanical analysis (TMA) measures material sample displacement as a function of temperature, time, and applied force. TMA is typically used to characterize thermal expansion, glass transitions, and softening points of materials.

thermoplastic resins. Polymers or copolymers that essentially have no cross-links and can be melted or reshaped when heated to melt point or above the T_g.

thermoplastic welding. The melding together of thermoplastic materials by taking the materials to a temperature above the melt point in order to weld them together.

thermoset resins. A plastic that cures by the application of heat or by chemical reaction resulting in a cross-linked insoluble molecule; considered an irreversible process.

three-dimensional (3D) fabrics. Fabrics that have structural fibers running in the Z-axis, generally between two plies of biaxial fabric.

through-transmission ultrasonic inspection. A non-destructive inspection technique where ultrasonic waves are sent from a transmitter to a receiver through the part (usually coupled with water). The signal is then processed and shown on a multi-color display so that the operator can determine if there are any reductions in the signal, indicating an anomaly.

tooling, tools. Molds, fixtures, jigs, templates, and other equipment used in the manufacture of other tools, parts, and assemblies.

tool sealer. A polymeric surface sealer used as part of a semi-permanent release system that is designed to fill micro-pores at the tool surface and minimize the possibility for mechanical attachment to the surface.

tool or mold surface. The primary surface of a composite part or assembly controlled by tooling for the purpose of aerodynamic smoothness, cosmetic or aesthetic requirements, or other necessity.

torsion load. The twisting stress applied to a structure.

toughness. The ability of a material to absorb shock or impact without damage.

tow. A unidirectional bundle of filaments for use in filament winding or fiber placement applications. Also used for manufacturing unidirectional tape.

tracers. Contrasting colored threads woven into fabric to identify the warp and fill directions; also used to show the fiber-dominant surfaces of harness satin weave fabrics.

transesterification. Conversion of an organic acid ester into another ester of that same acid. This conversion process can have a profound effect on certain cyanate ester resins.

transverse properties. Properties measured in the transverse direction, in-plane and perpendicular to the longitudinal direction of a laminate or structure.

triaxial reinforcement. A fiber form with three primary axes of fiber reinforcement.

trim fixture, *or* jig. Tooling used to hold and provide net-trim off-set dimensions, designed to be used for the purpose of trimming the component with a hand-held router or grit-edge rotary saw.

Twaron®. The trademarked name of an aramid fiber reinforcement used in manufacturing and repair operations.

twill weave. A weave style that allows for additional drapeability and conformance to complex geometric shapes.

twin sheet forming. A process in which two sheets of FRTP are processed at the same time in a dual-die mold-set, resulting in a completed, hollow-cored structure.

twisting in laminates. Twisting of an FRP laminate, in a post-processed laminate or panel, is typically due to an asymmetric layup. (See *symmetry*)

ultimate service temperature. The maximum temperature that a material can sustain for a short period of time with little to no degradation.

ultimate stiffness. The highest flexural or bending load that a material can sustain before failure.

ultimate strength. The highest tensile or compressive stress that a material will sustain before failure.

ultra high modulus (UHM) carbon fiber. Made from a pitch precursor, this type of fiber is processed at very high temperatures within an inert gas atmosphere to achieve the desired properties.

ultrasonic inspection/testing. A non-destructive inspection method using ultrasonic frequency sound waves to detect flaws in a structure.

ultrasonic leak detector. An ultrasonic receiver that "listens" to a high frequency band for noise emitted from a leak in a vacuum bag that may not be detectable to the human ear.

ultraviolet (UV) light, or radiation. Radiation from the ultraviolet light spectrum. UV light can degrade certain materials over time. Composite parts are usually painted with opaque colored paint to resist UV degradation.

unidirectional reinforcement. A fiber form or fabric where 80% or more of the fibers are running in one direction.

unidirectional tape. Dry or prepreg tows placed side by side, so that all of the structural fibers are running the length of the roll. Dry "tape" is usually held together with a zigzagging thread lightly bound to the backside of the tape with an inert binder. Prepreg tape is supported on a coated paper backing.

unreinforced plastics. Refers to a molded or cast thermoset resin or a thermoplastic sheet having no reinforcing fibers or structural fillers within the material.

unsaturated polyester. Condensation polymer formed by the reaction of polyols and polycarboxylic that contain double bonds, the typical polyester thermoset resin used in the composite manufacturing industry.

vacuum bag, vacuum bagging. The application of a flexible plastic or elastomeric membrane, sealed to the mold about the perimeter, from which vacuum is drawn to allow atmospheric pressure to be applied for the purpose of compacting and/or bonding materials together. Can be used in an autoclave to allow many atmospheres of pressure to be applied to a part or bonded assembly.

vacuum bag schedule, or sequence (of vacuum bag materials). Refers to the layers between the part and the vacuum bag, i.e. peel ply, release film, breather materials, etc. These materials can vary from part to part and are not always the same, particularly in non-autoclave, resin-bleed applications where the bleed is controlled to maintain resin content, yet allow for gas and air movement (with the resin) during processing.

vacuum bag sealant tape. Rubberized putty used to seal the vacuum bag film to the mold or fixture surface.

vacuum forming. A thermoforming process where a heated sheet of thermoplastic is draped over a vacuum form mold, pulled to the surface and into detail areas through a series of holes in the mold drawing vacuum.

vacuum gauge. A gauge that measures vacuum in millibars or inches of mercury.

vacuum leaks. Refers to leaks in a mold, fixture or vacuum bag during processing. Typically, there is specified leak rate that is deemed acceptable for processing.

vacuum port. A piece of hardware designed to penetrate either the vacuum bag or the mold/fixture, through which vacuum is drawn on the vacuum bag for processing.

vacuum pump. An air pump designed to pull vacuum at the inlet instead of pump air at the exhaust. A vacuum pump may be a portable unit, or a large pump attached to a tank and plumbing system for use by an entire facility.

Van der Wals force. A weak force of attraction between non-polar molecules caused by a temporary change in dipole moment due to a shift of orbital electrons from one side of one atom or molecule to the other, stimulating a similar shift in adjacent atoms or molecules and maintaining a weak attraction or bond. Named after Dutch physicist Johannes Diderik van der Waals (1837-1923).

vinyl ester. Vinyl esters are chemically similar to unsaturated polyester resins with the backbone of an epoxy resin. Like polyesters, they are relatively low cost and have similar thermal and mechanical properties but not necessarily the adhesion properties of an epoxy.

viscoelastic. The change in a thermoset resin from a liquid (visco) to a solid (elastic) state.

viscosity. The resistance of a material to flow; measured in Poise (P) or more commonly, centipoise (cP). The viscosity of water is a reference at 1 cP at 20°C/68°F.

vitrification. A solid or glassy phase of a material. When a thermoset vitrifies, it has become solid. (However it is not necessarily cured to the temperature capability desired.)

voids, void content. A void is a location in a laminate where there is no resin or fiber in that space. The void content is usually calculated based on mass to volume measurements. Refer to ASTM D 2734 for void volume calculation.

Volan finish. A Hexcel Schwebel proprietary surface finish used on glass fabrics to enhance wettability and bonding to polyester, vinyl ester, and epoxy matrices.

Volatile Organic Compound (VOC). Compound that has a high vapor pressure and low water solubility, typically controlled from direct release into the atmosphere for environmental and health concerns.

volatiles, volatile content. Materials such as water and alcohol, inherent to a resin formula that can be turned into vapor under vacuum, at ambient or low temperature, allowing for evacuation from the resin. The *volatile content* is the percentage of volatile that is contained in the resin.

volume ratio (or, fiber/resin ratio). Refers to the volume percentage ratio of fiber to resin contained in a composite laminate.

warp face. The face of a harness satin weave fabric that has predominately warp-yarns visible on the surface.

warp yarns, warp fiber direction. The longitudinal yarns in a roll of woven fabric; i.e., the yarns that run the length of the roll parallel to the selvedge. The primary yarns in a fabric used for axial orientation purposes.

weave, weaving, woven fabric. Any bidirectional fabric has been woven in a loom, with warp fibers alternating position at certain intervals in the weaving process so the fill yarns can be placed across the loom, thereby producing either a plain, twill, harness-satin, or other weave pattern.

weight ratio (or, fiber/resin ratio). Refers to the weight percentage ratio of fiber to resin contained in a composite laminate.

wet layup. A fabrication process where both freshly mixed resin and dry fiber forms are laid up in a mold or fixture.

wet-out. The action that takes place when a liquid (resin) is applied to a material (fiber) and the material either "wets out" or the liquid "beads up." This wetting action is crucial for good distribution of resin throughout the fiber form (tow, mat or fabric) and obtaining a good bond between the resin and fiber. A finish is usually applied to a fiber form or fabric to enhance wetting or wet-out.

wet T_g. The thermal glass transition temperature of a wet or water-saturated composite matrix.

working time. The actual working time of a mixed thermoset resin system, dependent upon the volume mixed and the amount of time left in mass in the bucket or cup while working, before spreading into a thinner cross-section. The actual time to gel in the final form.

woven roving. Fabrics with interwoven multiple yarns, bundled side by side (for bulk) and woven in a plain weave format. (*See* basket weave fabric)

yarn. A bundle of filaments or strands twisted together for use in a weaving process.

yarn count. The number of yarns per square inch in both the warp (W) direction and the fill (F) direction.

Zylon® fiber. A Hexcel Schwebel trademark name for their polyphenylenebenzobisoxazole (PBO) fiber. It is comparable to aramid in properties and does not hydrate at the same rate as aramid. It also can service a higher temperature range than aramid fibers.

Index

About the Authors

Louis C. Dorworth

Louis C. (Lou) Dorworth has been working in the composite manufacturing industry since 1978, with experience in research and development, manufacturing engineering, material and process engineering, and tool design and fabrication. His aerospace composite career began in 1979 with the Lear Fan 2100 program. He has since been involved with many other aircraft programs such as the Beech Starship 1, the B-2 bomber, the C-17 transport, the Airbus A-330/340, and Boeing 777, mostly working for first-tier sub-contractors. Lou has also been involved with many other non-aerospace programs ranging from wind blade manufacturing, to snowboards, to underwater submersibles.

Lou has been associated with Abaris since its inception in 1983. He has been teaching composite-related courses for Abaris part-time since 1989 and full-time since 1992. Currently Lou is the Division Manager of Abaris Direct Services, part of Abaris Training Resources, Inc., located in Reno, NV. In this position Lou is responsible for coordinating on-site training courses and technical consultation projects for Abaris' clientele.

Lou has been a professional member of The Society for the Advancement of Material & Process Engineering (SAMPE) since 1982. In 1994 he became a senior member of The Society of Manufacturing Engineers (SME), and is currently participating in their Plastics, Composites, and Coatings (PCC) Technical Community, as a primary member of the Composites Technical Committee (CTC).

Ginger L. Gardiner

Ginger Gardiner has a Mechanical Engineering degree from Rice University and has worked in the composites industry since 1990, when she was hired by DuPont as a technical marketing representantive for Kevlar® aramid fiber and Nomex® aramid honeycomb in aerospace and marine applications. She then formed a consulting company, performing market research and development for companies such as Hoechst Celanese, Ciba-Geigy, Abaris Training and Hexcel. She now lives in North Carolina with her husband, boatbuilder Jim Gardiner, and two children, and is a contributing writer for *High Performance Composites Magazine, Composites Technology Magazine* and *Professional BoatBuilder Magazine.*

Greg M. Mellema

Greg Mellema has been an Airframe & Powerplant mechanic since 1988. He holds an Inspection Authorization from the FAA and is an adjunct instructor of Aviation Maintenance Technology at Embry-Riddle Aeronautical University. Greg has nearly 30 years experience working on both civilian and military aircraft, building and repairing both metallic and advanced composite aircraft structures. During that time he has worked extensively with the U.S. Army's Test and Evaluation Command and now works full-time for the Army's Aviation and Missile Research Development and Engineering Center (AMRDEC) developing prototype advanced composite parts, training, and repair techniques for Army aviation. Greg holds a B.S. in Professional Aeronautics and a Masters degree in Aeronautical Science from Embry Riddle Aeronautical University and is a member of SAMPE and PAMA.

The Abaris Story

Founded in 1983, Abaris started as a small research and development company specializing in advanced composite materials and processes, mostly serving the aerospace composites industry. At the time, services included low-rate production of specialty components along with a single five-day training course on advanced composite structures fabrication and repair.

Spun from the former Lear Fan program, and headed by Mr. William L. (Bill) Murphy, a small group of former Lear Fan employees brought their specialized skills, knowledge, and expertise to Abaris, offering a variety of capabilities to the booming composites industry.

Within a few years, Abaris had ramped up their capabilities, moving from the old Lear Fan location at the Stead, NV airport to a bigger facility in nearby Reno, Nevada. They dedicated a classroom and lab area to training. In the mid-1980s, Abaris added a two-week course in advanced composite tooling, to complement the fabrication and repair training.

In 1988, Mr. Michael J. Hoke joined the Abaris staff and was tasked with overseeing the training side of the business. In 1989 Mr. Hoke and Mr. Murphy formed a business partnership and the training division was officially created. Upon the passing of Bill Murphy in 1991, Mike Hoke became the president and sole owner of Abaris Training Resources, Inc., a company dedicated solely to advanced composite training. Abaris was now serving a wider assortment of customers in industries including sporting goods, medical, motorsports, marine, civil engineering, wind energy, and aerospace, all of which are involved in the manufacture or repair of advanced composite structures.

In 1991 a new training facility was established in southeast Reno with a dedicated classroom, layup room, grinding/trim-room, and a large workshop area designed to meet the high demand for ongoing training in composite manufacturing, tooling, and repair technologies. Since then, Abaris has more than doubled the size of its Reno facility, allowing for two simultaneous courses to be taught at the same time at the Reno headquarters. The company has also added additional locations, one in Griffin, Georgia, and another co-located within the Lufthansa Resource Technical Training (LRTT) facility in the U.K., where it can better serve the European and Asian countries.

In addition to the facilities, Abaris has broadened its capabilities and now offers 20 different courses in engineering and design, repair design and analysis, manufacturing, adhesive bonding, fabrication and damage repair, bonded aluminum structural repair, rotor blade repair, wind blade repair, resin infusion, tooling, and nondestructive inspection (NDI). Abaris also offers on-site composite training at customer locations worldwide. Because of the high demand, in 2007 they formed the Abaris Direct Services Division. This division coordinates on-site training as well as customized training, engineering services, and consultation.

Abaris continues to expand along with the composites industry. There is a huge demand for knowledge in advanced composite technology, and it is the Abaris mission to stay on top of the latest developments and to impart this knowledge to the industry in a manner that is technically accurate, engaging and practical. This idea follows the Abaris goal to be the first choice for composite training.